The Systematics Association Special Volume Series 73

Biodiversity Databases

Techniques, Politics, and Applications

The Systematics Association Special Volume Series

Series Editor

Alan Warren
Department of Zoology, The Natural History Museum,
Cromwell Road, London SW7 5BD, UK.

The Systematics Association promotes all aspects of systematic biology by organizing conferences and workshops on key themes in systematics, publishing books and awarding modest grants in support of systematics research. Membership of the Association is open to internationally based professionals and amateurs with an interest in any branch of biology including palaeobiology. Members are entitled to attend conferences at discounted rates, to apply for grants and to receive the newsletters and mailed information; they also receive a generous discount on the purchase of all volumes produced by the Association.

The first of the Systematics Association's publications *The New Systematics* (1940) was a classic work edited by its then-president Sir Julian Huxley, that set out the problems facing general biologists in deciding which kinds of data would most effectively progress systematics. Since then, more than 70 volumes have been published, often in rapidly expanding areas of science where a modern synthesis is required.

The *modus operandi* of the Association is to encourage leading researchers to organize symposia that result in a multi-authored volume. In 1997 the Association organized the first of its international Biennial Conferences. This and subsequent Biennial Conferences, which are designed to provide for systematists of all kinds, included themed symposia that resulted in further publications. The Association also publishes volumes that are not specifically linked to meetings and encourages new publications in a broad range of systematics topics.

Anyone wishing to learn more about the Systematics Association and its publications should refer to our website at www.systass.org.

Other Systematics Association publications are listed after the index for this volume.

The Systematics Association Special Volume Series 73

Biodiversity Databases

Techniques, Politics, and Applications

Edited by

Gordon B. Curry

University of Glasgow
Glasgow, Scotland

Chris J. Humphries

The Natural History Museum
London, UK

CRC Press
Taylor & Francis Group
Boca Raton   London   New York

CRC Press is an imprint of the
Taylor & Francis Group, an **informa** business

CRC Press
Taylor & Francis Group
6000 Broken Sound Parkway NW, Suite 300
Boca Raton, FL 33487-2742

© 2007 by the Systematics Association
CRC Press is an imprint of Taylor & Francis Group, an informa business

No claim to original U.S. Government works

ISBN-13: 978-0-415-33290-3 (hbk)

Library of Congress Cataloging-in-Publication Data

Catalog record is available from the Library of Congress

**Visit the Taylor & Francis Web site at
http://www.taylorandfrancis.com and the**

**CRC Press Web site at
http://www.crcpress.com**

Contents

Chapter 9

Chapter 10

Preface

Since the first desktop computers emerged in the late 1970s and early 1980s, the power, speed and storage capacity has increased radically, especially in recent years. Indeed, the whole approach to computing and database management has shifted from the independent researcher keeping records for a particular project to state-of-the-art file storage systems, presentation and distribution over the World Wide Web. Taxonomists are natural information scientists and their outputs are highly desired by anyone interested in biology and geology, or indeed any system that requires retrieval of data. However, to make systems work effectively, it is necessary to bring together social scientists, programmers, database designers and information specialists to achieve the right political setting and give institutions the right platforms for dissemination of taxonomic information. This is what this book is about.

The subject matter is moving at a very fast pace; new techniques in ways of recognition, compilation and data management emerge virtually on a daily basis. The rules are changing and moving into a different league. The Intel and Motorola chips are very different processors compared with those of 10 years ago. Just as the World Wide Web gave access to vast amounts of information, the computing community is changing gear and raising the stakes with new capabilities for storage and moving information around. New initiatives are emerging that will bring together new agencies in the world of bioinformatics. Chapter 1 by Lane and Edwards and Chapter 2 by Los and Hof suggest that techniques of bioinformatics should be upgraded to levels achievable at global and European standards. Those chapters by Berendsohn and Geoffroy and Scoble and Berendsohn suggest that networking systems are trying to put the techniques together.

Within biology, computing is moving to a new generation in terms of function and phylo-informatics. Gone are the days of looking at one small group of organisms; molecular systematics, sequence storage and barcoding of all major groups of taxa mean that subjects such as blast searching, verification and delivery systems are changing the language of databases. Coupled with the notion that computers are far more useful now than they have ever been, this means that the time is right to bring together a group of professionals in the field.

Many of the databases dealt with are still handled by individual taxonomists, and these represent the cottage industry aspect of the task. Therefore, the purpose of this book is to show how we might turn the cottage industry into a major enterprise. At the same time, we acknowledge that the principles of database design were created by the pioneers and that what is needed is evolution rather than revolution in the field of development. At the end of the day, we want to be able to have access to the materials and methods without necessarily knowing where the original information comes from. This has been referred to as the industrialization of information amongst Australian colleagues. In Australia, there is a checklist of angiosperms online (a topic we had hoped to cover), but nobody really knows who is holding the record; we just know that it is agreed upon throughout Australia.

Other chapter authors in this volume have been at various cutting edges in their fields. For example, Andrew Jones and Richard White write about structures of databases, e-science, and their uses; Sterling et al. discuss analytical databases on conservation. Triebel et al. deal with the problems of Ascomycetes, and Curry and Connor write about automated extraction of database data from published descriptions. MacLeod et al. discuss species definitions and neural nets using computerized procedures, Jones presents new developments in computing, such as the grid, and White writes of linking databases together.

We have vision of a virtual biodiversity laboratory: validation, training of new systematists, utilization and empowering the new generation — all of whom will have access to the same, best, verified and accepted information. For example, drug plant Web sites will all have verified data to create confidence in their quality, with the view of eliminating erroneous records. Identity of species must be verified and linked to types and figured specimens; the importance of such repositories is that they hold the key to the names — the ultimate arbiters of good taxonomic identity. We hope that, by creating this book, we are not necessarily looking for answers to the big systematics questions of the day, but rather the means of getting through politically, socially and economically so that they can be delivered at the right levels through the Internet or whatever delivery vehicle is appropriate. New server protocols will change the architecture in such a way that there is a totally open-ended broadband. We see this book taking a step in an ongoing exercise that we hope will be repeated soon as the potential is more widely recognized.

Our aim is not to offer an inclusive view. We have to note that all future developments will be at the start of a new computer age, however good or bad they may be. There is a risk of proceeding without strong controls on the data presented; systematics protocols along with peer review offer the only guaranteed way of maintaining trust in the output.

We acknowledge support from the Linnean Society and the Systematics Assocation. Furthermore, we are indebted to our Irish hosts at the biennial symposium at Dublin University and the Council of the Systematics Association for continued support.

The Editors

Gordon B. Curry is currently reader in earth sciences in the Department of Geographical and Earth Sciences at the University of Glasgow, Scotland. Prior to joining the staff at the University of Glasgow in 1992, he was a Royal Society of London University research fellow for 8 years. His interests include palaeo-environmental reconstruction (in particular using stable isotopes), taxonomy and computing. He was project manager for the UK's Natural Environment Research Council's Centre of Research and Training in Taxonomy in the University of Glasgow for 5 years, until 1999. He acted as tutor for the Open University's evolution course from 1985 to 2003 and as associate lecturer of the university for the earth and life course in Scotland from 1996 to 2004.

Dr. Curry served on the Council of the Systematics Association from 1993 to 2005, and was treasurer from 1996 to 2005. He also served as the Systematics Association's representative on the Council of the Linnean Society, London, from 1999 to 2003. In 1985 Dr. Curry received the President's Award of the Geological Society of London and the Clough Award from the Edinburgh Geological Society. In 1989 he was awarded the Wollaston Fund from the Geological Society of London.

To date, Dr. Curry has prepared over 120 publications and written or edited five books. He has worked with seven postdoctoral research fellows and research assistants and supervised 19 Ph.D. projects, all successfully completed. Dr. Curry has organized nine international conferences in Scotland, England, France, Japan and Ireland. His research has been carried out primarily across Europe and in New Zealand.

Christopher J. Humphries is merit researcher at the Department of Botany at the Natural History Museum, London, and visiting professor at the University of Reading. He received his Ph.D. from the University of Reading. Dr. Humphries's interests are in systematic theory, angiosperms, historical biogeography and area selection techniques in conservation biology. Among the honours and awards that he has received are the Bicentenary Silver Medal 1980 (scientist of the year under 40 years), the Linnean Society of London's OPTIMA Silver Medal for 1979 for best paper on European taxonomy, 1979–1980, and the gold medal for botany awarded by the Linnean Society of London in 2001.

A few of the many positions held by Dr. Humphries include head curator of the European herbarium, British Museum (natural history), London, 1974–1980; principal scientific officer, general herbarium, British Museum (natural history), 1980–1990; and president of the Systematics Association, 2000–2003. He has served as associate editor of the *Botanical Journal of the Linnean Society*. Dr. Humphries was founder and editor of *Cladistics*, the journal of the Willi Hennig Society, and he is on the editorial board of the *Journal of Comparative Biology*.

Contributors

Walter G. Berendsohn
Botanic Garden and Botanical Museum
  Berlin–Dahlem
Freie Universität Berlin
Berlin, Germany

F. Borchsenius
Department of Biological Sciences
Ny Munkegade
Aarus, Denmark

Richard J. Connor
Department of Computer Science
University of Strathclyde
Glasgow, Scotland

Gordon B. Curry
Department of Geographical and
  Earth Sciences
University of Glasgow
Glasgow, Scotland

J. Dransfield
The Herbarium
Royal Botanic Gardens, Kew
Richmond, Surrey, UK

James L. Edwards
GBIF Secretariat
Copenhagen, Denmark

Marc Geoffroy
Botanic Garden and Botanical Museum
  Berlin–Dahlem
Freie Universität Berlin
Berlin, Germany

Cees H.J. Hof
Institute for Biodiversity and Ecosystem
  Dynamics and Zoological Museum
University of Amsterdam
Amsterdam, The Netherlands

Chris J. Humphries
Department of Botany
The Natural History Museum
London, UK

Andrew C. Jones
School of Computer Science
Cardiff University
Cardiff, Wales, UK

Meredith A. Lane
GBIF Secretariat
Copenhagen, Denmark

Wouter Los
Institute for Biodiversity and Ecosystem
  Dynamics and Zoological Museum
University of Amsterdam
Amsterdam, The Netherlands

Norman MacLeod
Department of Palaeontology
The Natural History Museum
London, UK

Thomas H. Nash III
School of Life Sciences
Arizona State University
Tempe, Arizona

M. O'Neill
Oxford University Museum of Natural
  History
Oxford, UK

Derek Peršoh
Universität Bayreuth
Bayreuth, Germany

Gerhard Rambold
Universität Bayreuth
Bayreuth, Germany

Malcolm J. Scoble
Department of Entomology
The Natural History Museum
London, UK

Ole Seberg
The Natural History Museum
  of Denmark
Botanical Garden and Museum
Copenhagen, Denmark

Jacob Andersen Sterling
The Natural History Museum of Denmark
Botanical Garden and Museum
Copenhagen, Denmark

Dagmar Triebel
Botanische Staatsammlung München
Department of Mycology
Munich, Germany

Steven A. Walsh
Department of Palaeontology
The Natural History Museum
London, UK

Richard J. White
School of Computer Science
Cardiff University
Cardiff, Wales, UK

Luciana Zedda
Universität Bayreuth
Bayreuth, Germany

1 The Global Biodiversity Information Facility (GBIF)

Meredith A. Lane and James L. Edwards

CONTENTS

ABSTRACT

In very broad strokes, as indicated by the International Union for the Conservation of Nature and Natural Resources (IUCN) in 1980, biology can be thought of at three levels of organization: molecular/genetic, species and ecosystem. The raw data of the molecular level are nearly all digital, as are many of those at the ecosystem level. However, the raw data of the species level (where they are found, the physiology, morphology, etc.) are almost all entirely analogue and descriptive. However, developments in informatics at each of these levels can be of service to the others. The Global Biodiversity Information Facility (GBIF) was established to enable the digital capture and dissemination of data related to natural history specimens (including those in culture and other living collections), of which there are an estimated 1.5 billion in at least 6000 collections worldwide. Another of GBIF's tasks is to generate an electronic catalogue of names of known organisms, which is the element required to enable data mining across all three levels in a single query. GBIF's work at the species and specimen levels of biological organization can be thought of as unifying the biological information domain. In addition, it provides worldwide coordination among the many ongoing digitization projects, standards development and networking efforts within biodiversity informatics.

1.1   WHAT IS GBIF?

The Global Biodiversity Information Facility has a mission to make the world's species' biodiversity data freely and universally available via the Internet. It is a megascience facility — in part because the GBIF concept was developed by a working group formed by the Mega-Science Forum (now the Global Science Forum) of the Organisation for Economic Cooperation and Development. More importantly, it is megascience because it is a worldwide endeavour that is challenging in the several areas of information science, technology and sociology as well as biology.

GBIF's efforts are focused on primary scientific biodiversity data at the specimen and species levels because these data, unlike most molecular/genetic and much ecological data,

1

are not in digital form. Nonetheless, primary biodiversity data of these types are critically important for society, science and a sustainable future.

The kinds of data and services that the activities of GBIF and its participants around the world will provide to the Web over the next few years include:

- georeferenced specimen data;
- an electronic index to scientific names and thus to the scientific literature and databases; and
- a means to link together data from disparate sources (e.g., DNA sequences, specimen illustrations, morphological characters, species observations and ecosystem data) to answer complex questions.

Among other things, georeferenced species occurrence data allow for

- better prediction of areas most suitable for wildlife reserves;
- rapid identification of, and information about, control of invasive species;
- prediction of patterns of spread of new diseases;
- correlation of species occurrence with ecological parameters and therefore the ability to understand effects of ecological change; and
- repatriation of biodiversity information to countries of origin.

Examples of the many sorts of applications and analyses that will be able to make use of these data include:

- systematic, taxonomic, ecological and environmental research;
- policy and decision making;
- natural resource management;
- conservation; and
- bioprospecting and biotechnology.

GBIF is a distributed facility, comprising a network of participant nodes that

- share biodiversity data openly and freely;
- use common standards for data and metadata;
- encourage generation of additional data;
- ensure that data providers retain control of their data; and
- share a common philosophy.

The GBIF philosophy is that primary scientific data should be available to all the different kinds of users, no matter where in the world they may be located. Analyses can be applied to the same data sets to answer different kinds of questions. By reusing data, duplication of effort is avoided. GBIF is also working towards the time when biological data and information from all levels of organization (molecular/genetic, species and ecosystems) can be interoperable and complex questions requiring information from all of those levels can be asked via single queries through a single Internet portal. Furthermore, different portals to the same data can be constructed, depending on the needs of particular users.

1.2 WHY WAS GBIF ESTABLISHED?

Calls from governments, industry and the public for biodiversity information have been increasing steadily because such basic information is needed for environmental decision making, scientific investigation and economic development. GBIF was established to make primary scientific data about natural history specimens and species occurrences available to everyone, no matter where in the world they live. Furthermore, biodiversity is unevenly distributed across the globe (with high numbers of species in the tropics, for instance). Likewise, biodiversity data are also unevenly distributed, but in this case predominantly in the developed countries of the temperate parts of the world. GBIF was established, in part, to redress the inequality of the distribution of the information by

- undertaking biodiversity informatics activities that must be accomplished on a worldwide basis to be fully useful;
- taking on tasks not being attempted by other initiatives but that would be of benefit to those initiatives (such as The Clearing House Mechanism of the Convention on Biological Diversity and The Global Taxonomic Initiative); and
- making biodiversity databases interoperable among themselves and with molecular, genetic, ecological and other types of databases, thus increasing the value of all.

1.3 THE GBIF CONTRIBUTION TO INTEROPERABILITY

GBIF's area of data and infrastructure development responsibility is unique. There is no duplication of any existing effort. GBIF is promoting the digitization of the label data on natural history specimens that have accumulated over the past 250–300 years, as well as the migration of observational data sets into modern information management systems and onto up-to-date platforms.

As shown in Table 1.1, the data within other segments of the biological information domain are already largely digital. Once GBIF has accomplished parts of its goals to 1) generate an electronic catalogue of the names of known organisms (ECAT) compilation of all scientific names, including their lexical and orthographic variants, that will function as a global electronic searching index and 2) to promote the digitization of natural history collections, linking databases from across the whole biological information domain will be

TABLE 1.1
The Biological Data Domain

Subdomain	Digital status	Data status	Greatest informatics problems
Molecular sequence and gene/genome data	95% Digital	Persistent digital data stores; universally accessible	Data migration, cleansing, vouchering, taxonomy (gene and species)
Species and specimen data	<5% Digital	Persistent physical data stores; accessible with difficulty	Digitization, migration of legacy data, indexing
Ecological and ecosystem data	80% (?) Digital	Persistent (?) digital and physical data stores; moderately accessible	Migration of legacy data, metadata generation, taxonomy (species)

possible. The return on the investments already made in the other areas will be enhanced by the data and interoperability provided by GBIF.

Many partners are working together to build a GBIF network that will serve science and society. These partners include all the GBIF participants (85 as of this document) and that number is growing all the time. Again, as of the writing of this chapter, some 115 million specimen records and more than one million scientific name records are available via the GBIF data portal (http://www.gbif.net). We anticipate that those numbers will also grow rapidly. GBIF welcomes all new potential partners in its endeavour to provide primary scientific information about specimens and species, as well as links to data and information from other levels of biological organization.

2 The European Network for Biodiversity Information

Wouter Los and Cees H.J. Hof

CONTENTS

ABSTRACT

Since the early 1990s, a rapidly expanding number of European projects have been initiated, all with the aim of organizing the appearance of biodiversity information in electronic databases. At the present time, the emphasis of these projects is on linking these databases together and on placing them in the framework of the Global Biodiversity Information Facility (GBIF). In order to create a common platform for these diverse projects, and to organize the European contribution to GBIF, the European Network for Biodiversity Information (ENBI) was established in 2003. ENBI will provide a centralized and clear overview of the interrelationships between all projects and initiatives and will promote a cooperative approach in support of the objectives of GBIF. ENBI is also identifying new plans and opportunities and supports some prioritized feasibility projects, with the aim of accelerating key aspects of the biodiversity infrastructure that are not yet in place. The combined efforts in ENBI are expected to provide a clear plan for how biodiversity resources should be maintained and developed in the twenty-first century.

2.1   INTRODUCTION

In comparison with the rest of the world, Europe contains a minor proportion of the Earth's total biodiversity. Europe is defined here as the biogeographic region from the North Pole down to and including the Mediterranean Sea, and from the Ural Mountains in the east to the Atlantic Ocean in the west, and also includes a number of islands in the Atlantic Ocean. However, as a result of the early development of taxonomy as a scientific discipline in Europe, this continent now curates about half of the world's biological collections. These collections comprise more than 50% of the described species and type specimens from all over the world. A significant number of internationally recognized taxonomists are also based in Europe, mostly working in one of the numerous natural history institutions. The largest of these institutes have organized themselves in the Consortium of European Taxonomic Facilities (CETAF [1]).

In order to provide better access to all available biodiversity information, a number of projects have been initiated to digitize and disseminate biodiversity data in all their formats. Both databases and complex information systems were developed on disk, on CD-ROM or as advanced online services. The relevant major European-wide projects are summarized in this chapter. With the growing number of databases and information systems, a new set of issues and problems emerged related to the need to integrate dissimilar data from different data owners and to provide customized functionalities to different user groups. Several projects address these issues for species databases, ecosystem databases and specimen databases. The Global Biodiversity Information Facility (GBIF [2]) triggered numerous developments and, for Europe specifically, the establishment of the European Network for Biodiversity Information (ENBI [3]).

2.2   PROJECTS THROUGHOUT EUROPE

Since the start of the present computer age, a wide variety of individuals and institutes across Europe started to exploit the newly emerging possibilities, concentrating their efforts on databasing, on digitizing taxonomic monographs and on preparing electronic identification keys. During the last decade of the twentieth century, a number of these initiatives developed into international cooperative projects. Crucial to these major projects were the so-called research framework programmes of the European Union, which created a number of opportunities to develop digital research infrastructures for biology. The taxonomic research community was amongst the first to submit coordinated proposals in order to establish biodiversity information services. A number of successful European-wide projects will be described in this chapter. The Web addresses of these projects are listed in the Cited WWW Resources section of this chapter.

2.2.1   SPECIES NAMES AND DESCRIPTIONS

Species name checklists have a central position in biodiversity information systems because they serve as the central directories leading to a wide range of digital information sources. In interaction with the international Species-2000 initiative, three projects on European species started to compile digital checklists. The first project benefited directly from the Framework Programme priority on marine ecosystems and led to the creation of the European Register of Marine Species on the Web (ERMS [4]). Subsequently, two other projects started with terrestrial and freshwater organisms. Euro+Med Plantbase [5] covers the vascular plant species, including the Mediterranean species of North Africa, while Fauna Europaea [6] tackles all multicellular animal species. In each of these projects, qualified expert taxonomists were selected to check the quality of the available species descriptions. The number of digitized species available is different for each project:

European Register of Marine Species	32,000
Euro+Med Plantbase	37,000
Fauna Europaea	130,000

Species-2000 Europe [7] started in 2003, with the aim of interlinking the three checklist databases into a single European gateway, thereby contributing directly to the Global Biodiversity Information Facility.

Turning to the much more detailed information available in species descriptions, the Europe-based Expert Centre for Taxonomic Identification (ETI [8]) cooperates with experts worldwide to build fully digital monographs on various groups of organisms. These monographs include advanced multiple-entry identification keys and distribution data. Initially, the monographs were published on CD-ROM, but they are now also partially accessible via the Internet. Other cooperative projects have been working on a variety of Web-based information systems for specific taxonomic groups or in relation to a specific topic.

2.2.2 Collection Specimen and Observation Data

Biological collections of primary importance for biodiversity research include those housed in natural history museums and herbaria, botanical and zoological gardens, microbial and tissue culture collections, and plant and animal genetic resource collections, as well as the observation databases (surveys, mapping projects). Europe houses the most extensive living and natural history collections as well as survey data collections of global importance. Taken together, this represents an immense knowledge base on global biodiversity.

In a series of projects, different institutes across Europe have come together to develop and implement a Biological Collection Access Service for Europe (BioCASE [9]). The BioCASE project provides standardized metadata, taking into account the complex and changing scientific (taxonomy, ecology, palaeontology) and political/historical (geography) concepts involved. BioCASE also enables user-friendly access to the specimen information contained in biological collections (see Chapter 4).

Special kinds of collections data are available for micro-organisms. In 1998, the Organisation for Economic Cooperation and Development (OECD) decided to identify so-called (microbial) biological resources centres (BRCs) that would act as key information components of the scientific and technological infrastructure of the life sciences and biotechnology. BRCs would consist of the service providers and the repositories of living cells, genomes and all information relating to heredity and the functions of biological systems. More specifically, BRCs contain collections of culturable organisms (e.g., micro-organisms and cells from plants, animals and human), replicable parts of these (e.g., genomes, plasmids, viruses, cDNAs), viable but not culturable organisms, cells and tissues, as well as the databases with molecular, physiological and structural information relevant to these collections and related bioinformatics. Several European initiatives did contribute to this process, becoming a BRC with an emphasis on data services, such as the Microbial Information

Network Europe Project (MINE), the Common Access to Biological Resources and Information project (CABRI [10]) and the more recently created European Biological Resources Centres Network (EBRCN [11]).

2.2.3 PLANT GENETIC RESOURCES

As is the case with genetic sequence databases, biodiversity databases in this area are primarily focused on cultivated plants. These resources are also addressed in the Convention on Biological Diversity, and all countries are therefore obliged to create national inventories of plant genetic resources (PGRs). The European Plant Genetic Resources Information Infra Structure (EPGRIS [12]) aims to establish an infrastructure for information on PGR maintained *ex situ* in Europe by (1) supporting the creation of and providing technical support to national PGR inventories; and (2) creating a European PGR search catalogue with passport data on *ex situ* collections maintained in Europe. The catalogue is frequently updated from the national PGR inventories and is meant to be accessible via the Internet. This European inventory will be called EURISCO (European Internet Search Catalogue, a name derived from the ancient Greek word meaning 'I find') and it will automatically receive data from the national inventories. It will effectively provide access to all *ex situ* PGR information in Europe and thus facilitate locating and accessing PGRs. The project will support countries in this task through workshops, technical advice and staff exchanges and by developing standards.

2.2.4 DNA AND PROTEIN SEQUENCES

The European Molecular Biology Laboratory maintains the EMBL Nucleotide Sequence Database (also known as EMBL-Bank [13,14]), which is Europe's primary nucleotide sequence resource. The main sources for DNA and RNA sequences are the direct submissions from individual researchers, submissions from major genome sequencing projects and patent applications. The database is produced in an international collaboration with GenBank (USA [15]) and the DNA Database of Japan (DDBJ [16]). Each of the three groups collects a portion of the total sequence data reported worldwide, and all new and updated database entries are exchanged between the groups on a daily basis.

As a supporting network, EMBnet has evolved, during its 15 years of existence, from an informal network of individuals in charge of maintaining biological databases into a network organization bringing bioinformatics professionals together to serve the expanding fields of genetics and molecular biology. EMBnet nodes provide their national scientific community with access to high-performance computing resources, specialized databanks and up to date software. Many nodes act as redistribution centres for national research institutes. In addition, staff from several EMBnet nodes collaborate in developing new biocomputing tools and to give specialized courses at their nodes.

An important recent development is a large subsidy from the European Commission to 24 bioinformatics groups based in 14 countries throughout Europe to create a pan-European BioSapiens Network of Excellence in Bioinformatics. The network aims to address the current fragmentation of European bioinformatics by creating a virtual research institute and by organizing a European school for training in bioinformatics. A common goal of these developments is to overcome the data overload, which is reaching epidemic proportions among molecular biologists. The network will coordinate and focus excellent research in bioinformatics by creating a virtual institute for genome annotation. Annotation is the process by which features of the genes or proteins stored in a database are extracted from other sources and then defined and interpreted. The institute will also establish a permanent European school of bioinformatics to train bioinformaticians and to encourage best practice in the exploitation of genome annotation data for biologists.

2.2.4.1   Ecosystem Data

Ecosystem data are difficult to deal with because any data presentation assumes that it is possible to classify ecosystems in discrete elements that can be represented in standardized databases. Cooperation throughout Europe contributed to the European Vegetation Survey (EVS), with the intention to develop common data standards, computerized databases with portable software and a standardized classification of plant communities. In contrast, the European Union CORINE [17] Biotope Classification provides a catalogue of habitats and vegetation, but it has few data on biodiversity. The EUNIS [18] habitat classification has been developed to facilitate harmonized description and collection of data across Europe through the use of criteria for habitat identification. It is a comprehensive pan-European system, covering all types of habitats from natural to artificial and from terrestrial to freshwater to marine habitats.

A new development following from the preceding was the SynBioSys (Syntaxonomic Biological System [19]) project. This project developed a computer program to classify ecological communities above the species level, but now in relation to the species composition in such communities. The system works on two levels: plant communities and landscapes. The plant community level is based on data with respect to species composition, ecology, succession, distribution and nature management. An interesting application of this resource is that it provides an identification system that allows users to assess which plant communities best fit with their own observed data. A digital vegetation database with data compositions from the years 1930–2000 serves as the basis for this system. For the landscape level data, physical geographic regions are also included in the database.

2.3   START OF THE EUROPEAN NETWORK FOR BIODIVERSITY INFORMATION

ENBI [3] was established in January 2003, following a call from the European Commission to better organize and network all European activities that may contribute to the goals of GBIF [2]. As such, ENBI has the general objective of managing an open network of relevant biodiversity information centers established in the western European pale-arctic region. ENBI includes all European national GBIF nodes and all relevant EU-funded projects. Other important stakeholders are also represented, and altogether, ENBI hosts over 60 institutes established in 24 countries. ENBI operates as a network, so the emphasis is on interaction between all partners in order to identify, prioritize and test (potential) new developments through a number of e-conferences, workshops and feasibility studies. Because ENBI operates in close cooperation with GBIF, the work plan priorities are in many respects similar to those of GBIF. However, ENBI also explores other new developments as a potential contribution to future GBIF efforts. The work plan of ENBI is organized in four main clusters.

2.3.1 COORDINATING ACTIVITIES

The first cluster coordinates all activities in order to establish a strong biodiversity information network. Strategies for sustainability and continuity should be supported by a common European, or preferably a global, approach. Critical questions being addressed by this cluster include which activities and digital services should be organized locally or internationally and whether these services should be provided in the public or in the private domain. The partnership in ENBI has to address these problems in order to get a view on the future landscape of all activities in biodiversity information and informatics. This includes the difficult issues relating to intellectual and ownership rights of digital data in a shared Web environment. A realistic opinion on which activities will continue to require a common approach and are more efficiently managed at the European scale will provide the basis for a business plan to be discussed with the relevant European authorities.

In this cluster, another important task deals with the dissemination of expertise, especially with regards to the training of new generations of biodiversity informatics specialists. The network organizes a number of workshops in different parts of Europe, and it is hoped that it will also influence plans for curriculum development at universities.

2.3.2 MAINTENANCE, ENHANCEMENT AND PRESENTATION OF BIODIVERSITY DATABASES

The second cluster deals with common approaches for the development, enhancement and maintenance of databases with taxonomic, specimen, collection and survey data. This should result in the promulgation of the rational use of techniques, including best practice policies. An example is the Global Lepidoptera Names Index [20] to which ENBI contributed financially in order to develop recommended approaches, which were then distributed throughout Europe. Another example is a workshop on techniques and challenges for digital imaging of biological type specimens. Network partners are cooperating to identify gaps in knowledge and information, to accelerate databasing and to develop appropriate strategies. A main problem for all database custodians is the presently insufficient routines and mechanisms to update, validate and ensure sustainability of the databases. In interaction with the previously mentioned specific European projects, the ENBI partners are looking for generalized solutions so that the various networks and institutions can efficiently share and reuse information without duplication of efforts.

2.3.3 DATA INTEGRATION, INTEROPERABILITY AND ANALYSIS

The third cluster in ENBI is investigating general options for the integration and interoperability of large-scale distributed databases (genetic, species, specimen and ecological), together with relevant information from other domains such as chemical compounds, geography, climate or economic activity. By making inventories of analytical software systems, the network hopes to promote new technologies to utilize the wealth of growing biodiversity databases. New opportunities exploiting the potential of Grid developments are of particular interest. Interoperability between the heterogeneous data systems and common access to all biodiversity information will create the opportunity to perform analysis on the large amount of European data available. Analytical tools are mostly installed within single biodiversity information systems. However, a number of initiatives include Web-based analytical tools based on a variety of distributed databases. ENBI will focus on

GIS in biodiversity analytical systems as a model for further development in specific (for example, national) applications.

2.3.4 USER NEEDS: PRODUCTS AND E-SERVICES

The last cluster in ENBI aims to provide mechanisms that will support the development of communication platforms to meet end-user priorities with respect to high-quality products and e-services. In the European context of different languages, it would be an important service to users if they had access to information in their own languages. ENBI is making dictionaries of biodiversity terminology in a number of European languages, which can be integrated in existing machine translation services. In another network activity, partners are cooperating to find the best procedures to serve specific users' needs that require the involvement of different, and changing, data providers. The (semi-automatic) provision of custom-made services will require much attention because user requests (such as on policy issues) mostly require difficult solutions.

Requests that can be handled are not restricted to European data. Europe holds the world's richest and most important biodiversity collections, literature and other data and much of this information relates to parts of the world other than Europe; thus, the network will also contribute information to users outside Europe. By sharing data with GBIF, the network hopes to accelerate the success of GBIF.

2.4 PARTNERS IN THE NETWORK

The contributing partner institutes in the network have been identified as coordinating institutes of past and current European projects in biodiversity information or informatics or as designated GBIF nodes. In total there are more than 60 partners involved. Because many partner institutes coordinate specific networks, the whole ENBI network is effectively much larger. A smaller number of institutes have been identified to take a leading task for the various task clusters and more specific work packages in ENBI. Together, they constitute the steering committee responsible for overseeing the progress of the network activities. A Memorandum of Understanding, in collaboration with the European Environment Agency, has been established to define the contributions from each participating organization.

CITED WWW RESOURCES

1. CETAF (Consortium of European Taxonomic Facilities): http://www.cetaf.org/
2. GBIF (Global Biodiversity Information Facility): http://www.gbif.org and http://www.gbif.net
3. ENBI (European Network for Biodiversity Information): http://www.enbi.info/
4. ERMS (European Register of Marine Species): http://erms.biol.soton.ac.uk/
5. Euro+Med Plantbase: http://www.euromed.org.uk/
6. Fauna Europaea: http://www.faunaeur.org
7. Species-2000 Europe: http://sp2000europa.org
8. ETI Biodiversity Center: http://www.eti.uva.nl
9. BioCASE (Biological Collection Access Service for Europe): http://www.biocase.org/
10. CABRI (Common Access to Biological Resources and Information): http://www.cabri.org/
11. EBRCN (European Biological Resource Centres Network): http://www.ebrcn.org
12. EPGRIS (European Plant Genetic Resources Information Infra Structure): http://www.ecpgr.cgiar.org/epgris/

13. EMBL Nucleotide Sequence Database: http://www.ebi.ac.uk/embl/index.html
14. EMBNet (European Molecular Biology Network): http://www.embnet.org/
15. GenBank: http://www.ncbi.nlm.nih.gov/
16. DNA Database of Japan: http://www.ddbj.nig.ac.jp/
17. CORINE (land cover database): http://terrestrial.eionet.eu.int/CLC2000
18. EUNIS European Nature Information System: http://eunis.eea.eu.int/index.jsp
19. SynBioSys: http://www.synbiosys.alterra.nl/turboveg/
20. Global Lepidoptera Names Index: http://www.nhm.ac.uk/entomology/lepindex/

OTHER USEFUL SITES

EU DataGrid project: http://eu-datagrid.web.cern.ch/eu-datagrid/
Global Grid Forum: http://www.gridforum.org/
TDWG (Taxonomic Databases Working Group): http://www.tdwg.org/

3 Networking Taxonomic Concepts — Uniting without 'Unitary-ism'

Walter G. Berendsohn and Marc Geoffroy

CONTENTS

3.1 INTRODUCTION

One of the principal aims of current efforts in biodiversity informatics is to network the electronically available information about organisms from a wide variety of sources. This information has been produced at different times and places and with differing aims and is normally pigeonholed by means of the organism's scientific name. However, correct (accepted) names are formed according to rules of nomenclature, without regard to the concept or circumscription of the taxon itself (Berendsohn 1995). Potentially, correct names stand for differing concepts (potential taxa). Consequently, names are not providing a reliable index for biodiversity information, but electronic networks such as the Global Biodiversity Information Facility or the European BioCASE (see Chapter 4, this volume) do need such an index for information access.

That taxonomic concepts pose a problem for taxonomic databases was recognized already a decade ago (e.g., Beach et al. 1993). In contrast to information provision mediated by individuals (normally specialists), public databases need integrated explicit knowledge to reliably transmit complex information. In the 1990s, information models laid the theoretical base for handling taxonomic concepts — names used in the sense of a certain circumscription (Zhong et al. 1996; Berendsohn 1997; Le Renard 2000; Pullan et al. 2000; Ytow et al. 2001; Anonymous 2003). Later, software also demonstrated the practicability of such models for taxonomic data (e.g., Pullan et al. 2000; Gradstein 2001). This article summarizes and updates results from our working group published earlier (Berendsohn 2003), presents further evidence for the relevance of the problem, and reports progress made with the implementation of the Berlin Model (Berendsohn et al. 2003).

Many still doubt that there is a need for the representation of concepts in taxonomic databases. For technical implementation, it is certainly easier to do without such a feature. Even taxonomists often tend to think only of the basic scientific aspect of their endeavour; that is, the product of their work — the latest taxonomic treatment — should be regarded as the state of the art and earlier works should stand corrected. However, this disregards the fact that names of organisms are widely used by non-taxonomists and have been used so for a considerable amount of time. To ask questions such as 'How stable are concepts in taxonomy?' and 'How reliable are taxonomic names as an index to biodiversity information?' is not just a hobby of the information modellers' community, but rather an essential requirement for the entire systematics community.

For many groups, we do not know the answers to these questions; in some groups, we know that there is considerable stability (especially if supported by the nomenclatural methodology (e.g., in bacteriology), and in others we suspect that there is a high degree of instability, but we lack hard data. Explicit statements of concept conflicts (and lack thereof!) are rare and mostly hidden in monographer's notes, etc. Two recent publications, both for groups of plants, provide data to assess concept stability and thus the extent of the problem.

3.2   ASSESSING CONCEPT STABILITY

For an analysis of concept stability, we need more data than are contained in traditional checklists with their lists of synonyms (although these can also be used for analysis; see Geoffroy and Berendsohn 2003). Instead, we need data sets where concept relationships have been recorded in a comprehensive way (i.e., for every taxon processed) rather than in the largely anecdotal way found in traditional treatments. In analysing such data sets, we have to be conscious of the fact that the resulting values for stability strongly depend on three factors: (1) the selection of other works with which the current concept is compared; (2) the quality of the comparison itself; and (3) the degree of taxonomic creativity of the authors of the present work itself.

While neither selection nor quality can be assessed in a quantitative way, the third factor, here called 'novelty', can be assessed. We simply calculate the amount or percentage of concepts for which no congruent concept is cited in the existing literature used for the comparison. A low novelty value indicates a conservative treatment, which will yield values that depend less on the current treatment and more on the variability of concepts already in use.

The first example publication is the *Standard List of the Ferns and Flowering Plants of Germany* (Wisskirchen and Haeupler 1998), which was and is maintained as a database at the German Federal Agency for Nature Protection (BfN 2004) and available on the World Wide Web via the agency's Flora Web portal. In this publication, 4709 accepted taxa are listed; 3811 of these are species and the rest are infraspecific taxa. The novel feature of this work is that Wisskirchen and Haeupler indicate the relationship of their taxon concept with that in a number of contemporary floristic works commonly used by students and practitioners of the German flora to determine plant species (or properties of plant species). Figure 3.1 gives an excerpt to illustrate the data.

This work was clearly not carried out with the aim of analysing concept stability, and it only states congruence or non-congruence of concepts without finer details. Nevertheless, the fact that it is fully databased gave us the opportunity to attempt an analysis of the data

ANAGALLIS L. (Primulaceae) – Sp. Pl.: 148 (1.5.1753) – Typus:
Anagallis arvensis L. – NCU – <u>Gauchheil</u>

    CENTUNCULUS L. – Sp. Pl.: 116 (1.5.1753) – Typus: Centunculus minimus L. – NCU

Anagallis arvensis L. – Sp. Pl.: 148 (1753)* – Typus: Herb. Linn. No. 208.1, LINN (lecto-   E F h M O R S
DYER in DYER et al. 1963: Fl. Southern Africa 26: 14)
    <u>Acker-Gauchheil</u>

        Anagallis caerulea L. – Amoen. Acad. 4: 479 (1759)
        Anagallis phoenicea SCOP. – Fl. Carniol., ed. 2, 1: 139 (1771)*, nom. illeg. (nom. superfl.)
        Anagallis carnea SCHRANK – Baier. Fl. 1: 461 (1789)
        Anagallis arvensis subsp. phoenicea VOLLM. – Ber. Bayer. Bot. Ges. 9: 44 (1904), nom. inval.   H
        Anagallis arvensis fo. carnea (SCHRANK) LÜDI   O
        Anagallis arvensis var. phoenicea GOUAN
        Anagallis arvensis var. caerulea (L.) GOUAN – Fl. Monsp.: 30 (1765)
        Anagallis arvensis fo. azurea HYL. – Uppsala Univ. Årsskr. 7: 256 (1945)   O

   Im Gebiet nur die subsp. *arvensis*

Anagallis foemina MILL. – Gard. Dict., ed. 8: no. 2 (1768)   E F M O R S
    <u>Blauer Gauchheil</u>

        Anagallis caerulea SCHREB. – Spic. Fl. Lips.: 5 (1771) non L. 1759, nom. illeg.

FIGURE 3.1 An extract from Wisskirchen and Haeupler. The main body of text is a typical botanical checklist, with the correct (accepted) names in boldface and an abbreviated citation of the publication of the name and of the type specimen. The common name is followed by a list of synonyms (i.e., names that have the same type or names the type of which is considered to be included in the concept the correct name stands for). The unusual feature of the work is embedded in the letter codes on the right, which stand for floristic works in current use in Germany. Uppercase letters indicate that the name is considered to be used with the same circumscription in those works; lowercase letters indicate a different concept. Missing letters indicate that the name was not used. For *Anagallis arvensis*, six works use the same name with same concept, and one has a different concept. For *Anagallis foemina*, six works use same name with same concept and one does not use name or the concept. (From Wisskirchen, R. and Haeupler, H., eds., *Standardliste der Farn- und Blütenpflanzen Deutschlands*. Ulmer, 1998. Reproduced with permission of Verlag Eugen Ulmer.)

set. To start with the assessment of novelty, we think the numbers indicate that a conservative approach was taken: 62% of the taxon concepts in the list are also treated in *all* other works and, for 93% of the taxon concepts, at least one congruent concept is cited among the other works included in the comparison. As to concept and nomenclatural stability, the data set confirms that roughly half of German vascular plant taxa are stable for name *and* concept and 60% are stable for concept, throughout the works compared.

These results are in good agreement with the data available for German mosses in a study much more focused on the concept issue. For their *Reference List of the Mosses of Germany*, Koperski et al. (2000) used the concept-oriented IOPI model (Berendsohn 1997) for their data recording. They related the 1548 accepted taxa to concepts in 11 floristic or taxonomic treatments, which they considered to be in current use (Figure 3.2). Koperski et al. (2000) based their assessment on detailed comparison of descriptions and discussion of the concepts. The relationships between potential taxa (PT) documented represent the five basic relationships between two concepts that can be stated when the concept or potential taxon is perceived as a set of objects (specimens, observations, etc.): (1) PT1 and PT2 are congruent; (2) PT1 is included in PT2; (3) PT1 includes PT2; (4) PT1 and PT2 overlap each other; and (5) PT1 and PT2 exclude each other.

Of the 1548 accepted potential taxa (PT) in Koperski et al. (2000), 1509 (97%) have one or more congruent concepts within the compared works. A low degree of novelty can be recognized.

Dicranum fuscescens Sm.
Fl. Brit. 1804 ‹27›

= *Dicranum congestum* Brid.

≙	*Dicranum fuscescens* Sm.	sec. CORLEY & al. (1981/1991)
≙	*Dicranum fuscescens* Sm.	sec. LUDWIG & al. (1996)
	LUDWIG & al. (s. dort, S. 289) berufen sich auf das Konzept von CORLEY & al.	
≙	*Dicranum fuscescens* var. *eu-fuscescens* Mönk.	sec. MÖNKEMEYER (1927)
<	*Dicranum fuscescens* Turner	sec. FRAHM & FREY (1992)
	incl. *D. flexicaule* (siehe Anmerkung dort)	
<	*Dicranum fuscescens* Turner	sec. MÖNKEMEYER (1927)
	incl. *D. flexicaule* (siehe dort)	
<	*Dicranum fuscescens* Sm.	sec. SMITH (1980)
	incl. *D. flexicaule*, das in der typischen Varietät enthalten ist (vgl. morphologische Beschreibung)	
>	*Dicranum fuscescens* var. *congestum* (Brid.) Husn.	sec. SMITH (1980)
	Bei diesem Taxon handelt es sich offensichtlich um eine montane Wuchsform von *D. fuscescens*.	
≶	*Dicranum fuscescens* var. *fuscescens*	sec. SMITH (1980)

FIGURE 3.2   An extract from Koperski et al. The top three lines represent a traditional checklist entry, with the correct name, its protologue citation (with an indication of the source), and a list of synonyms (here, only one; the equal sign indicates that it is a heterotypical synonym). The following lines cite several potential taxa (i.e., names accepted in a certain reference [the citation following the 'sec.' for secundum, according to]). The symbol in the beginning shows the concept relationship with respect to the accepted (correct) name: congruent, included in, including and overlapping. (Koperski, M. et al., *Schriftenreihe Vegetationsk*, 34, 1–519, 2000. Reproduced with permission of the German Federal Agency for Nature Protection.)

In terms of concept stability, wider concepts have been found among the other treatments for 515 (33%) of the taxa listed, while included concepts were present for 267 taxa (17%), and overlapping concepts were identified for 90 taxa (6%). Including the data that can be extracted from the traditional synonymic list, we come to the following result of the analysis of the moss data set: with respect to nomenclatural and concept stability, 55% of the taxa show stability with respect to concepts; that is, they cite no relations to other concepts but congruent ones. These 55% can be further subdivided. For 35%, we can state that only homotypical synonyms have been cited, so there remains little doubt about their stability. However, only a small percentage of all taxa listed offer that level of stability under a constant name. For 20%, there is some indication of instability (e.g., heterotypical synonyms or misapplied names are cited). As opposed to that, 45% of the taxa show explicit instability; that is, concept relationships other than congruent ones are cited (Figure 3.3).

In summary, therefore, there is a high degree of instability of concepts in a considerable proportion of plant taxa, even among works in current use. Content linked to taxon names includes, *inter alia*, uses (mostly human) and threats (to species, to hosts, to health, to environment, etc.), ecology (pollination, symbiosis, parasitism, indicator value, edaphic and climatic requirements, etc.) of species, molecular data (natural substances, genes, sequences, physiology, etc.), geographical range or occurrence and descriptive data. Kirschner and Kaplan (2002) have strikingly demonstrated how compilation of lists of names can lead to inaccurate information of high practical importance (in that case, red lists of threatened

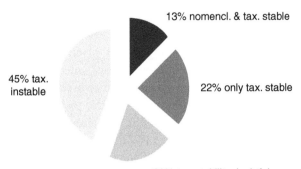

FIGURE 3.3 Nomenclatural and taxonomic stability in German mosses. (Data from Koperski, M. et al., *Schriftenreihe Vegetationsk*, 34, 1–519, 2000. Reproduced with permission of the German Federal Agency for Nature Protection.)

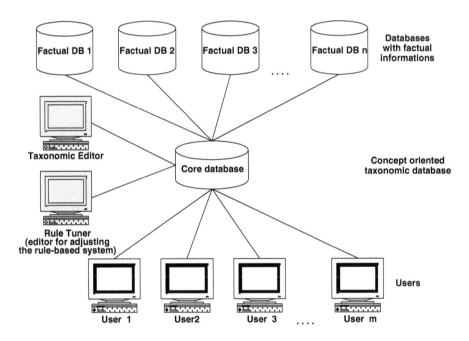

FIGURE 3.4 A simple model for access to information linked to taxon names.

plants). Considering the increasing ease with which these data can be linked using the Internet and considering the obvious hazards of uncritical linking of such knowledge by means of taxon names, systematists have to take action to construct systems that more reliably inform users about the caveats (or lack thereof) of information linkage.

3.3 A CONCEPT-ORIENTED SYSTEM — CURRENT STATE

Figure 3.4 depicts a system that relays information from providers of taxon-linked factual information to users querying on taxon names. This simple model can work given one of the following two scenarios: (1) all data providers agree on common concepts for the taxa involved; or (2) each provider is only dealing with a single taxonomic group, for which he or she provides the authoritative view. The second scenario was adopted by the Species 2000 system (Roskov and Bisby 2004), and the first scenario is supposedly followed by the ITIS system in the USA (ITIS 2004). We suggest that, given the inherent problems in taxon concepts and naming, this system is suboptimal at least when legacy data, divergent views and general resource discovery are to be supported. The broker module must be supported by a series of additional modules that effectively mediate the information access via names and allow a dependable transmission of factual information.

The central component of such a system is a database (which may be distributed over several sites) that allows the storage of taxon concepts and their relationships. Such a system will allow the linkage of different information sources independent of the concept and the transmission of information along the concept relationships established in the database. It will also allow calculations of the dependability of a name as the representation of a concept and thus permits conjecture on the concept stability for a name introduced without explicit concept relation (as is mostly the case for the factual databases).

Of course, the content of this database has to be edited and kept up to date, so an editor software component is necessary. Ideally, there should be one system for local data maintenance, drawing on fast connections and the functionalities available in a local client-server environment, and a separate remote editor system, allowing the database to be edited over the Internet.

Finally, this database also needs output tools for print media and for the World Wide Web, supporting the tasks the editor of a taxonomic work would need and hence encouraging taxonomists to actually use and improve upon the data in the database.

With the Berlin Model database (Berendsohn et al. 2003) and related tools, the essential components of database, editor and output tools have been put in place. The development has been supported by several projects, which share the core database model and functionality.

The EU-funded Euro+Med Plantbase project supported the development of the Internet editor software (Güntsch 2003). Another EU project, Species 2000europa, currently supports the installation of Berlin Model database and tools at the Euro+Med central site. The same project will also create a standardized access to the Euro+Med database and the IOPI database hosted in Berlin, which is also based on the Berlin Model. The IOPI database provides access to legacy data sets as well as checklist access to the taxonomic treatments published in the species *Plantarum* series. For example, the treatment of Juncaceae (Kirschner 2002) is published there in its original form; at the same time, a parallel data set is further improved and added to by the Juncaceae working group using the remote editor. Med-Checklist (vol. 2, Compositae) and the Dendroflora of El Salvador are two Berlin-based checklist projects currently using the database.

The AlgaTerra project has been instrumental in developing the local editor software and devising a comprehensive extension of the system to cover type specimens and their assertion. AlgaTerra involves cooperation among seven partners in Germany and links diverse information from molecular investigations, herbarium specimens and cultured strains via a common taxonomic core on micro-algae (AlgaTerra 2004). The same approach is also followed by the German Federal Agency for Nature Protection. As mentioned earlier, the German standard list is available online (BfN 2004); however, up to now it has only been available as a single-concept checklist. For obvious reasons, the users of taxonomic information have been driving the development of concept-based checklists in Germany. The agency primarily deals with information linked to names as opposed to the taxonomy itself. The standard list database is currently being converted to a Berlin Model database by the

MoReTax project. This system will also be used as the taxonomic access system for information on German specimens and observations within the German GBIF-Node for Botany (GBIF-D 2004).

In technical terms, the databases are currently implemented under MS SQL-Server (and Oracle), with cross-project coordination of database-level functions and procedures. The remote taxonomic editor was implemented using the ColdFusion application server and Java. The local taxonomic and extensions editor is based on Visual Basic, while database maintenance tools (data integrity checking mechanisms, etc.) as well as the WWW output use various clients and tools.

3.4 MAKING IT WORK

A database (acting as the taxonomic core), editor software to input and change data and database maintenance tools are available and already allow us to produce and publish traditional as well as concept-oriented checklists. We are now starting to meet the challenge to create a broker system incorporating the concept relationships present in a Berlin Model database, which acts as the system's taxonomic core.

The user may issue a query to get information about a certain taxon (name) from different sources (e.g., distribution information from one database, medical uses from another and red list status from a third). Equally, the user may directly query the content (red-listed organisms with medical properties from Germany). In both cases, taxon names are used to produce the result; the second case only differs in that the names to be searched for are coming from the content databases.

These databases may specify a taxon concept as their taxonomic reference point or only a taxon name. In the former case, matters are greatly simplified because the content can be directly linked to a concept in the taxonomic core. The following account of the broker's function will be based on the latter, currently prevailing case.

The broker performs the following functions:

1. It searches the taxonomic core database to retrieve all known names for the taxon.
2. It gets the requested content linked to these names from the connected databases.
3. It provides the content to the user, including statements to explain the way it has expanded the query in step 1, as well as caveats resulting from the taxonomic core's knowledge about concept instabilities for that particular name. This presentation of content strongly depends on the level of expertise of the user, which should be defined to at least distinguish taxonomists from the rest of the world.

The broker should provide as much trustworthy information as possible to the user. This may be simple in the case where the taxonomic core provides reasonable proof that all used names stand only for a single concept (all concepts are congruent; all synonyms are 'unequivocal' in Species 2000 terminology). However, as we have shown, this is not always the case even for a single specified name. Moreover, in many cases we still lack explicit statements as to concept relationships, and we have to rely on implicit information, such as that given in the taxonomic hierarchy (a subspecies is included in its species) or lists of synonyms (homotypical synonyms at least share their type, so their relationship is at least overlapping).

The broker has to rely on a transmission engine, a component responsible for selecting those concepts and names that are related to the given name and to which content information may have been linked. To disclose these relationships, even where they have not been explicitly stated, rules must be established that define, on the one hand, how far the engine should go in its processing — that is, to what depth the chain of possible consecutive relationships should be investigated (i.e., from PT_1 to PT_2, from PT_2 to PT_3,..., PT_{n-1} to PT_n). Alternatively, rules define the relationship arising between PT_1 and PT_n according to the particular relationships involved in the chain between them and the nature of the information to be transmitted (see following). Further rules of the transmission engine specify which information should be displayed to the user and which caveats should be listed, depending on:

- the resulting relationships to the concept to which this content was linked;
- the level of expertise of the user who issued the query; and
- the nature of the information to be transmitted.

For example, some information relates to every element in a taxon ('is a tree'), some to some elements ('has wings' in a taxon where larvae are wingless), and some to the entire set but not to individual elements ('occurring in Germany and France'). Such classes of information require different processing in the transmission engine and different displays.

In conclusion, to provide meaningful output, the system must consider a complex set of rules and parameters for the construction and use of relationships between taxonomic concepts. It also needs to know about the nature of the information transmitted and the level of expertise of the user of the system. This information has to be stored as part of the broker's transmission engine component, and an editing tool (the rule tuner) must be implemented to be able to tweak the output of the system.

The theoretical base for these components was detailed by Geoffroy and Berendsohn in several articles in Berendsohn (2003). Presently, we are starting to meet the challenge to create such a transmission engine and the rule tuner. The combination of the concept-based taxonomic core database (Berlin Model) and the transmission engine will help us to network and better utilize the growing number of available content providers for biodiversity information. First attempts to implement user interfaces with some of the transmission engine and rule tuner features are currently under way in the MoReTax and GBIF-D Botany projects, using the German plant data sets, and in the AlgaTerra project.

3.5 CONCLUSIONS

Users demand a Web-based unitary taxonomy (Godfray 2002) to get reliable access to species information. However, the taxonomic revision is as a rule not possible because local treatments, lack of new treatments or different hypotheses lead to co-existing taxonomies (Scoble 2004). Using modern IT tools, taxonomists can easily provide information on concept relationships between different systems and treatments, thus creating a pathway between current and past treatments. At the very least, specialists should make an effort to state where there appears to be no problem. Transmission models will allow using concept relationships — also those extracted from traditional synonyms and (perhaps) specimens — for an access system that relates information from different sources to the user. A concept-based taxonomic information system thus unites the taxonomic research process with reliable name-based user access to biodiversity information. Fleiureihe vegetiansk = Vegetelianskeunde or = Vegetianske.

ACKNOWLEDGMENTS

The team in Berlin working on the software components mentioned also includes Andrea Hahn, Anton Güntsch, Javier de la Torre, Jinling Li, Karl Glück, Markus Döring and Wolf-Henning Kusber.

REFERENCES

AlgaTerra (2004) AlgaTerra — An information system for terrestrial algal biodiversity: A synthesis of taxonomic, molecular and ecological information (http://www.algaterra.net).

Anonymous (2003) VegBank Taxonomic Data Models. Ecological Society of America (http://vegbank.org/vegbank/design/planttaxaoverview.html).

Beach, J.H., Pramanik, S., and Beaman, J.H. (1993) Hierarchic taxonomic databases. In *Advances in computer methods for systematic biology: Artificial intelligence, databases, computer vision*, ed. R. Fortuner. Johns Hopkins University Press, Baltimore, MD, 241–256.

Berendsohn, W.G. (1995) The concept of 'potential taxa' in databases. *Taxon* 44: 207–212.

Berendsohn, W.G. (1997) A taxonomic information model for botanical databases: The IOPI model. *Taxon* 46: 283–309.

Berendsohn, W.G. (2003) MoReTax — Handling factual information linked to taxonomic concepts in biology. *Schriftenreihe Vegetationsk* 39: 1–115.

Berendsohn, W.G., Döring, M., Geoffroy, M., Glück, K., Güntsch, A., Hahn, A., Kusber, W.-H., Li, J.-L., Röpert, D., and Specht, F. (2003) The Berlin Taxonomic Information Model. *Schriftenreihe Vegetationsk* 39: 15–42.

BfN (2004) [Mar 11]. *Bundesamt für Naturschutz [Federal Agency for Nature Protection]. Floraweb — Daten und Informationen zu Wildpflanzen und zur Vegetation von Deutschlan* (www.floraweb.de).

GBIF-D (2004) [Mar 11]. *The GBIF Programme in Germany (http://www.gbif.de/homeenglish)*.

Geoffroy, M. and Berendsohn, W.G. (2003) The concept problem in taxonomy: Importance, components, approaches. *Schriftenreihe Vegetationsk* 39: 5–13.

Godfray, H.C.J. (2002) Challenges for taxonomy. *Nature* 417: 18–19.

Gradstein, S.R., Sauer, M., Braun, M., Koperski, M., and Ludwig, G. (2001) TaxLink, a program for computer-assisted documentation of different circumscriptions of biological taxa. *Taxon* 50: 1075–1084.

Güntsch, A., Geoffroy, M., Döring, M., Glück, K., Li, J.-L., Röpert, D., Specht, F., and Berendsohn, W.G. (2003) *The taxonomic editor. Schriftenreihe Vegetationsk* 39: 43–56.

ITIS (2004) [Mar 4]. Integrated Taxonomic Information System online database. About ITIS — Background information (http://www.itis.usda.gov/info.html).

Kirschner, J., ed. (2002) *Juncaceae 1–3. Species* Plantarum: *Flora of the world part 6–8*. Australian Biological Studies, Canberra.

Kirschner, J. and Kaplan, M. (2002) Taxonomic monographs in relation to global red lists. *Taxon* 51: 155–158.

Koperski, M., Sauer, M., Braun, W., and Gradstein, S.R. (2000) *Referenzliste der Moose Deutschlands. Schriftenreihe Vegetationsk* 34: 1–519.

Le Renard, J. (2000) TAXIS, a taxonomic information system for managing large biological collections. In *Abstracts. TDWG 2000: Digitizing Biological Collections*. Taxonomic Databases Working Group, 16th Annual Meeting, Frankfurt, November 10–12, 2000, p. 18.

Pullan, M.R., Watson, M.F., Kennedy, J.B., Raguenaud, C., and Hyam, R. (2000) The Prometheus Taxonomic Model: A practical approach to representing multiple classifications. *Taxon* 49: 55–75.

Roskov, Y. and Bisby, F. (2004) [11 Mar]. Species 2000: An architecture and strategy for creating the catalogue of the world's plants (http://www.sp2000.org/presentations.html).

Scoble, M.J. (2004) Unitary or unified taxonomy? In *Taxonomy for the 21st century*, ed. H.C.J. Godfray and S. Knapp. *Philosophical transactions of the Royal Society (biological sciences)*, 359 (1444): 699–710.

Wisskirchen, R. and Haeupler, H., eds. (1998) *Standardliste der Farn- und Blütenpflanzen Deutschlands*. Ulmer.

Ytow, N., Morse, D.R., and Roberts, D.M. (2001) Nomencurator: A nomenclatural history model to handle multiple taxonomic views. *Biological Journal of the Linnean Society* 73: 81–98.

Zhong, Y., Jung, S., Pramanik, S., and Beaman, J.H. (1996) Data model and comparison and query methods for interacting classifications in a taxonomic database. *Taxon* 45: 223–241.

4 Networking Biological Collections Databases
Building a European Infrastructure

Malcolm J. Scoble and Walter G. Berendsohn

CONTENTS

ABSTRACT

Over recent years, several initiatives on improving access to information in natural science collections have been supported by the European Commission of the European Union. All are founded on the principle that the databases containing this information are scattered across numerous individual sites and servers, making the task one of constructing an integrated yet notably distributed network. This paper summarizes the history of this evolving exercise and examines its progress. Several issues are considered: the core task of connecting databases to the network is deeply influenced by the construction of the user interface; linking databases that are not entirely uniform in structure creates technical demands. No less demanding are problems of user access and the control of data quality.

4.1 INTRODUCTION

It is unsurprising that taxonomy has become engaged so intimately with computing. Its methods and protocols may be complex (demonstrably so in the Codes of Nomenclature), but they are logical and thus amenable to being modelled. The field is rich in data, with a literature spanning a period of close to 250 years, the nomenclatural basis of which (the binomial system) has continued unbroken. Our taxonomic system is founded on collections of specimens numbering, it has been suggested (L. Speers, pers. comm.), between 1.5 and 3 billion. This huge, but fragmented, resource is scattered across the globe in museums, herbariums, seed- and tissue banks and laboratories holding cultures of micro-organisms. Furthermore, the vast quantity of label data associated with the specimens remains largely trapped in the institutions housing collections, thus restricting access to it to those who can visit or borrow material (the latter involving risky shipment). Digitization of label data and the creation of digital images of specimens allow, potentially, users to gain electronic access to a wealth of information. Even greater in quantity than label data is the enormous number of observational records derived, particularly from survey work. Many data exist already, but there is the potential to gather far more.

The sum of these points underlies the observation that taxonomy is a *distributed* system. Not only is its raw material (collections, associated archives and literature and the holding databases) widely spread, but also, consequently, is the human expertise — the taxonomists who use and are typically associated with the physical resource. A similar situation pertains to biological recording. Consolidation of collections sometimes occurs, but nation states usually prefer to keep collections within their national or state boundaries. While facilities are distributed in most areas of science, they are particularly so in taxonomy; there are, for example, far more collections than particle accelerators. If taxonomic resources and expertise are so markedly distributed, so then is the best (if biased) sample of global biodiversity available to us.

To deliver effective access to this information, two integrated networks are necessary. One is a technical network by which databases may be linked and their collective content searched by means of a suitable user interface. The other is a network of people capable of creating, improving, protecting and sustaining the system. In the present chapter, we consider the position of database development and Internet access to information stored in databases within institutions housing natural history collections in Europe. Connections among three projects, all funded by the European Commission, lie at the heart of this endeavour. They include:

- a biological collections information service in Europe (BioCISE [1]);
- the European Natural History Specimen Information Network (ENHSIN [2]); and
- a biological collections access service for Europe (BioCASE [3]).

Closely related to these initiatives is the European Network of Biological Information (ENBI [4]), a specified task of which is to expand the BioCASE network of databases. Efforts to sustain the BioCASE infrastructure are also to be addressed in the networking activities of SYNTHESYS [5], an initiative of CETAF (the Consortium of European Taxonomic Facilities), which includes 19 European taxonomic facilities (natural history museums, botanic gardens and culture collections) funded under the Integrated Infrastructure Initiative of the European Commission's Framework Programme 6. Furthermore, several European country nodes of the Global Biodiversity Information Facility (GBIF) actively complement the BioCASE network.

The relationship between the projects (BioCISE and ENHSIN have been completed) has been complementary and evolutionary. Briefly stated, in BioCISE, knowledge about information structures in collections was consolidated, European data sources (collections and their holding institutions) were reviewed, and descriptive data about the collections (collections metadata) were gathered. A prototype system for connecting databases holding specimen (unit-level) data was developed in ENHSIN, and issues associated with collection networks and data accessibility over the Internet were explored. Seven European partners were involved.

In BioCASE, a pan-European operational system is being developed that unites and links collection- and unit-level data. A sophisticated user interface linked to a thesaurus to enable rich user searching is being developed. Thirty-five institutions from 30 European countries and Israel are taking part in the project. Establishing a strong European network for biodiversity information is the primary purpose of ENBI, with its role to contribute an integrated European dimension to GBIF. A networking activity within the Integrated Infrastructure Initiative in SYNTHESYS was organized to take BioCASE-related activities further beyond the end of the actual BioCASE project period.

4.2   BIOCISE: IDENTIFYING THE RESOURCE

During the course of their careers, specialist users of collections (notably taxonomists) build a working knowledge of the location and importance of those biological collections holding material relevant to their research. Their efforts could be much more effectively achieved if the data in this massive and fragmented resource were rendered more accessible as a true infrastructure. This observation applies to an even greater extent to other, less specialist users, who have a much more limited understanding of the resources available, and particularly to those barely, as yet, engaged with information-rich collections institutions.

BioCISE (Berendsohn 2000) was formed to survey the biological collections in the EU and consider the means of providing computer-based access to information contained within them. A detailed model of collection information completed under the BioCISE project was described by Berendsohn et al. (1999). This model illustrated the complexity of such information, yet demonstrated that because different biological subdisciplines share a similar information structure, they have the potential to make information available through a common access system.

In the BioCISE survey, it was estimated that over 4000 biological collections in public domain institutions might potentially contribute to a European data service (Hahn 2000). Furthermore, a clear message was received from many holders of collections that they were willing to share unit-level (specimen and observational) data. The huge number of existing biological specimens and observations makes the task of databasing at the unit level daunting. This is true not only for the process of keyboarding data on labels and preparing digital images of specimens, but also because of the need for data cleansing. The Internet has the power to disseminate misinformation (e.g., misidentifications, incorrect label data) much more rapidly and widely than does the paper medium, which is slower to circulate and where the content typically is refereed. In practical terms, it is impossible to imagine achieving comprehensive and accurate databasing of unit-level data in the foreseeable future. Fortunately, access to data content in biological collections can be eased through the provisions of collection-level data (Berendsohn et al. 2000).

Such data are descriptive of collections or subsections of collections. If such meta-information can be made accessible through a suitable interface, users can source data;

this allows them to understand such matters as what is available for study, in which collection, and when and by whom it was collected. Further advantages of this approach are that collection-level databases that lack homogeneity can be linked and data quality can be indicated. In many cases, collection-level information is the very material that users seek. Given the low percentage of specimens recorded in databases, it offers a practical solution to gaining access to a rich source of detail (Berendsohn et al. 1999).

Certainly the problems of achieving a clean set of specimen-level data with a significant coverage of what is available in collections-rich institutions are immense. Yet there is every reason to encourage the process of facilitation. A specifically European effort to this end has been made in two projects: ENHSIN and BioCASE. ENHSIN developed a prototype pilot network. BioCASE is expanding ENHSIN and combining it with the collection-level access of BioCISE.

4.3   PROVIDING ACCESS TO UNIT-LEVEL DATA: THE ENHSIN PROTOTYPE

At the heart of the European Natural History Specimen Information Network lies the technical development of a simple XML-based prototype for providing a common access system to specimen databases (Güntsch 2003). The system was designed to handle distributed and heterogeneous collections' databases. Projects outside Europe of a similar scope include Species Analyst [6] and REMIB [7]. The partnership that constituted the network also examined broader questions of the usage of such information, data quality, intellectual property and effective procedures to sustain networks of this kind after their creation.

The demonstrator access system has four major components (Güntsch 2003): the data sources, a user interface, a central XML client and the XML wrapper that is placed on each source database (Figure 4.1). Although some heterogeneity in databases is always likely to occur, even with the promotion of content standardization, information can at least be made

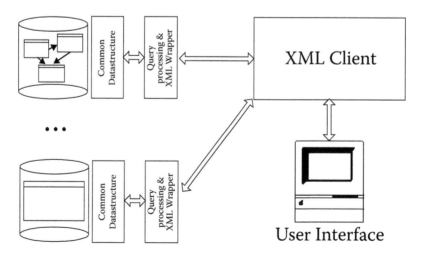

FIGURE 4.1   ENHSIN: system architecture. (From Güntsch, A. In *ENHSIN: The European Natural History Information Network*, ed. M.J. Scoble. Natural History Museum, London, 33–40, 2003 (http://www.nhm.ac.uk/science/rco/enhsin/publication/ENHSIN_ch_03.pdf))

available through unified data structure. Most potential data providers are only likely to engage with a network if the effort of taking part is minimal. Hence, it was essential for the ENHSIN pilot system to be designed in such a way as to demonstrate that databases with less structured fields could be accommodated. This approach meant that expectations will always be placed on the network developer to design functions and algorithms to handle heterogeneity.

Thus, an important feature of the ENHSIN pilot system was to demonstrate its tolerance to variation in data specification — its capability to cope with structured and unstructured data (Güntsch 2003). A well-structured record in a database can be exemplified by clear divisions (atomization) of the collecting site into fields of, for example, country, locality, latitude and longitude, and a division of the date of collection into day, month and year. By contrast, in a poorly structured record, the collecting site and date might take the form of a descriptive, unsegregated text. There are many intermediate situations between these extremes. While the atomized data may be of a wider immediate use (e.g., for mapping records), the unstructured record may still contain very valuable data that are useful for other purposes.

The ENHSIN user interface provides a query form with fields for genus, species, name of collector, date of collection, and country. The maximum number of records to be returned is specified by the user and an option is given to select all records, those for plants alone, or those for animals alone. Opting for fuzzy retrieval allows the user to search in less structured fields besides those that are well structured. The XML wrapper is installed on the Web server of the site of each data source. It enables queries to be processed from the XML client. The client receives queries from the user interface for transmission to the data sources and returns answers to the user as an HTML table.

To encourage data holders to engage with a collections network it is essential, as Güntsch (2003) explained, to design software that renders the process of installing the wrapper on a target database as simple an operation as possible. In ENHSIN, both unit- and collection-level elements (34 in total) describe the set of collection objects — known as the 'element set' (see http://www.bgbm.fu-berlin.de/BioDivInf/projects/ENHSIN/PilotImplementation.htm for the full list). This element set was defined for proof-of-concept purposes only. It was not proposed as a standard and it was not considered that such a simple set can satisfy the demands of user and provider communities for a rich data set that also covers issues such as intellectual property rights. In the pilot system, the data provider had first to portray information in the database by arranging data items in an order so that, at the very least, it contained mandatory elements. Creation of this portrayal or view enabled the wrapper software to be installed on the Web server of the data provider. The new information source was then registered by completion of a simple questionnaire, which solicited basic information about the collection. This was transmitted to the central system maintenance function.

As a demonstrator, the ENHSIN pilot system linked just seven specimen databases to the network. But the longer term intention was to expand the network across Europe and to include observation data, thus increasing vastly the number of databases linked in a biological collection access service. Such an expansion increases demands made on the software. The ENHSIN system requires data providers to use Microsoft server technology. To allow other Web services to be adopted, generic scripting language is needed to enable wrapper software to be implemented. A means of speeding the return of XML documents from the data sources to the central client software is also essential. The ENHSIN system relies on retrieval of results sequentially. While a sequential mechanism functions adequately when only a few databases are linked, it is too slow for a larger system to deliver information effectively. Therefore, parallel processes are required for a large network.

A further restriction of ENHSIN is that the means by which queries are transmitted to the XML wrappers from the central client are insufficiently flexible to make a network of databases most useful. A mechanism more appropriate for a fully operational network is the use of a standardized query language so as to locate and filter the data fields. Finally, the architecture of the ENHSIN system has the capacity to allow processing of a much wider element set than in the demonstrator, so it is intrinsically suitable for a wider network. These issues have been discussed by Berendsohn (2003 and elsewhere).

Besides the technicalities, several significant problems have to be understood and resolved if truly effective access to the wealth of data in or associated with collections-rich institutions is to be made. The three most demanding are a blend of the technical and organizational. The first is to enable users to perform deep searches of the network of databases so that simple inputs lead to rich outputs of interconnected data. The second is to ensure that the functionality of the system is of sufficient value to users: is it sufficiently reliable, of relevance to user needs, capable of being corrected and added to by means of suitably usable mechanisms? The third is finding suitable means of sustaining systems once they become operational.

Intellectual property issues are complex and need consideration in sustainability. Although the problems should in no way be underestimated, progress on actually posting data on the Web does not seem to have been of a sufficient magnitude to impede progress in BioCISE, ENHSIN and BioCASE or indeed other international projects.

Computerization of collections has been undertaken within institutions over many years for purposes of internal management. However, the Internet provides access to a wide range of users. Just a small fraction of information in natural history collections on biodiversity has been digitized. Typically, this information comes from specimens, card indexes and manuscripts. There is so much more to be added from existing and, particularly, future data sources that will result from recording schemes and surveys.

Yet the problems to be overcome are substantial (see Bailly 2003). For example, while it is a fundamental aim of this initiative to ease access to information, some data, such as those relating to the location of endangered species, need to be restricted. Furthermore, since data are not all of equal quality, their degree of access might be worth limiting. Specimens may be misidentified, localities may be imprecise, or miscoding of localities might occur in numeric coding schemes. Access can only be provided free at the point of use if the sustainability of the system is ensured.

The success of such a complex endeavour will stand or fall not only on the quality and extent of the data, but also on ease of access provided by the user interface and its underlying mechanism. Interface issues were explored in the ENHSIN project by Bailly (2003). Users have expectations: they must be able to search by different criteria such as taxon, geographic area, locality and institution. This means that a high-quality search system is needed. A simple input box may be adequate for the needs of specialists when making a specific search, but for the non-specialist user, browsing is likely to be more common. The ability to access data in one's own language can be extremely important — a matter of significance in Europe where there are so many tongues. Indeed, data may have been entered in different languages, so several terms may be used for a single country.

The prodigious quantities of information available, or likely to become available, in biological collection databases will require sophisticated filtering to prevent users from becoming swamped with or, conversely, deprived of key data. While standard desktop computer software does have data-sharing facilities, it is inadequate for the demands of the emerging task. A further problem with taxa and with geographical areas is that boundaries are often fuzzy. Plains and oceans, for example, typically have no boundaries and there are no standards on which to rely for data consistency.

The magnitude of these problems is significant. Yet progress towards solving some of them is being made. Although they may never be universally adopted, data standards are being encouraged, so they help provide a small part of the solution. An important means for users to search databases is a thesaurus, a complex example of which (drawn from the work of Copp, 2003) is outlined later for the BioCASE project. A significant development has been the widespread adoption of extensible markup language (XML), which allows data to be shared across the Internet by creating common information formats. Besides encouraging best practice in compiling databases, automated data-cleansing systems are being developed to improve quality. Data quality may be improved through feedback by users, and such feedback can be encouraged by rendering the mechanism effective and ensuring that responses are added to the databases.

The two most significant issues to be addressed on the broader management of the emerging access service are those of sustainability and the protection of intellectual property. Major collections-holding institutions would almost certainly have to take responsibility for sustainability. It remains unclear how hard-pressed collections managers will find the time to care for the collections under their charge, digitize further information, and manage a potentially massively expanding data source with user feedback increasing in line with this expansion and with the improved interfaces being developed. It is also unclear how the increased cost implications will be met, but if there is no new money and if charging users for data access is not a realistic option (which it almost certainly is not, other than in some highly specialist commercial situations), then a shift in staff working patterns and skills seems inevitable. What is unquestionable is that the collections community surely cannot avoid engagement with the sea change occurring in the development of the virtual collection that is fast developing alongside the physical collections.

Two main intellectual property issues have emerged during the development of the European collections access consortium, although there are several smaller ones. One is the very sensitive issue of providing data access to specimens collected from a country other than the one in which they are currently housed. While networks are predicated on the principle of enabling data to be shared, some parties, particularly developing countries, consider that it is the source country that should decide whether data should be shared and, if it is agreed that it should, then just what data. Nations with international collections are reluctant to 'repatriate' specimens. Data sharing is surely a better approach, one that can benefit all concerned. Attempting to restrict access to biodiversity data through claims of ownership of the intellectual property will be counterproductive to the entire biodiversity information effort. Resolution of the problem is likely to lie more within the domain of sociology than law.

The other issue reflects a duality within the minds of many staff in collections-rich institutes — notably museums and herbariums. While most curators and researchers are keen to share data in principle, there is often a residual concern that open access to an institution's data means a loss of quality control and a release of raw material for others to forge and create intellectual assets. Essentially, individuals and institutions need clear acknowledgement of the magnitude and significance of their input in terms of research effort or data access to relatively unstudied material. In the plural funded environment in which nearly all collections-rich institutions function today, managers are inevitably hopeful that their intellectual and material resources might sometimes generate funding, even though, with collections work, this has turned out to be extremely limited in practice. This matter has by no means been resolved. Part of the answer, at least in Europe, would be for individuals to make a mental shift from focusing on being employees of a particular institution to becoming part of the pan-European research area and the much wider research and infrastructure community that it represents.

The achievements of ENHSIN were thus technical (the pilot network) and exploratory (issues of user access, data quality, sustainability, intellectual property). But, importantly, the project also united partners who have continued as key players in the larger BioCASE project and, indeed, in other projects funded by the European Commission.

4.4 MAKING THE NETWORK OPERATIONAL WITH BIOCASE

While BioCISE made an assessment of biological collections across Europe and ENHSIN provided a prototype for a specimen-level network of databases, the task of taking these initiatives forward to form a pan-European, operational network falls under the much larger and ambitious initiative BioCASE [3]. Thirty-five partners across 31 countries take part in BioCASE. BioCASE combines in a single system a means of providing access to unit- and collection-level databases.

In Figure 4.2, the main components of the BioCASE system are shown. These are the central BioCASE core, the data sources (lower section of the figure) and the user interface (upper section), which is served by the thesaurus and indexing mechanism. Unit-level data are delivered to the core directly by means of a wrapper placed on the server of the data provider. Collection-level metadata are transmitted through a system of national nodes. Each node is responsible for describing biological collections in its particular country, and a facility exists to gather meta-information from special interest groups that cross national boundaries.

This simple summary belies the technical complexity of the project and the managerial demands of meshing the efforts of many collaborators. Delivery of the project requires achievement in several areas. Data interchange standards for data in collections need specification and the information flow requires organization. An extensible thesaurus of terms has to be developed for indexing the information and allowing users to query the network. To create an effective access service, a user-friendly interface is essential, and a careful analysis of user needs is necessary to inform the design of the user interface. Sustaining the system requires an appropriate business model, a part of the project closely linked with matters relating to the view that data providers have of their intellectual property and the degree of access that they are comfortable providing.

4.4.1 NATIONAL NODES AND COLLECTION-LEVEL DATA

An underlying principle of BioCASE and related projects is that data content should remain in the hands of the provider. Gathering the collection-level data is the responsibility of each

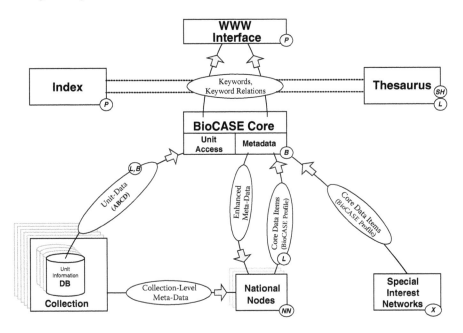

FIGURE 4.2   Flow chart of the BioCASE system. (Courtesy of the BioCASE secretariat.)

national node. Many nodes are national museums, botanical gardens or biological institutes and, as such, are bodies housing large collections of specimens and data. Each node hosts the national metadatabase, the content of which is acquired from the data providers through a Web site provided by the node. Wrappers placed on these national node databases enable them to be polled by the BioCASE central node in order to update the information held in the central access system, the core metadatabase (CorM).

National metadatabases contain, minimally, information describing biological collections in their respective countries. Content includes the names and descriptions of the organizations and individual collections, their address, the name of the contact person in each and the URLs. Examples of information specific to collections are the kind of collection, the nature of the objects and their number, and the number of species represented. Particular nodes may operate well beyond provision of the basic data, and they may play a role in identifying unit-level databases.

4.4.2   UNIT-LEVEL DATA

The diagram in Figure 4.2 shows that unit-level data are defined by the ABCD (access to biological collections data) schema (ABCD 2002). The ABCD schema design provides a data definition appropriate for all kinds of biological collections. It reaches far beyond a minimal common denominator approach and thus supports a wide range of database structures and data content. The detailed data profile enables collection holders to provide fully the rich data they hold on the individual units (specimens and observation records). Connection of variously structured databases to the BioCASE network is enabled by means of a CGI/XML wrapper placed on a given database that has been configured by the data provider. The protocol defining the technical communication between the different parts of the unit-level network system is called the BioCASe protocol — the small 'e' indicating its reach beyond Europe as the protocol for ABCD standard data provision in the GBIF (GBIF 2003).

4.4.3 A SEAMLESS SYSTEM

Because of the hierarchical nature of the meta-information describing biological collections (from details of an organization holding collections, their collections and subcollections down to the very specimens), access to collection- and unit-level data is possible as part of the same seamless system (Berendsohn et al. 2000). This is a major advantage, for if unit-level data are not yet available in digital form, users can at least gain access to collection-level data. In practice, providing users with easy access to the information is complex because of the broad scope of the information offered and the lack of clear definitions for some important elements of the meta-information, which may often change their concept considerably through time (e.g., locality boundaries and taxon concepts).

4.4.4 USER INTERFACE, THESAURUS AND INDEXING

These three components of the BioCASE system are strongly integrated. User access to data in the distributed system depends on an indexing module and a thesaurus. Data providers usually describe the content of their collections in the form of free text or keywords; therefore, to make that information accessible to users by means of structured searches, the indexing module will segregate the information provided and relate it to terms in the thesaurus. Development of the BioCASE thesaurus was described by Copp (2003), who noted that the common problem with any database is that, once it is constructed, it is difficult to see what it contains. In large relational databases — the situation applying to many in the domain of biological collections — the magnitude of the problem is considerable. Simple indexes as a means of searching are inadequate since their terms often have alternatives and these alternatives may have different contexts. Thus, the BioCASE thesaurus is best understood as a means of enabling data retrieval.

The thesaurus, which is in an advanced state of development, is being constructed by C.J.T. Copp (e.g., Copp 2003), who has also developed the complex data model for its management. Key features are:

- that it should be extensible by addition and by refinement (notably through linkage of terms to allow efficient and comprehensive searching;
- that its content will be derived from many sources: an expectation is that not only will the current (historical) base be incorporated, but so, potentially, will a vastly greater body of information that includes terms from various languages, a range of disciplines and alternative search terms;
- that it will be possible to capture and incorporate terms from meta-information about collections and by means of those entered by users searching the system;
- that the context of any term contained is capable of being preserved so that results from queries can be maximized or, for specific user groups, focused. For users to get the best value, terms must be placed in a meaningful context of other terms (e.g., 'insect' and 'invertebrate' need to be linked); and
- that its purpose is to facilitate searching rather than achieving accuracy, authority and completeness.

In summary, development includes the evaluation of existing data catalogues, followed by the creation of an extendible thesaurus (mainly taxonomic and geo-ecological), allowing for adequate representation of overlapping concepts as well as hierarchical, uncertain, imprecise and incomplete data (see BioCASE Web site [9]).

To deliver these ambitions is immensely demanding. Emerging metadata standards and their gradual incorporation in biological collection databases certainly help the task, but many sources remain idiosyncratic or at least exhibit considerable variation. Linking the wealth of potential thesaurus terms being gathered from public domain sources and from dictionaries created *de novo* is complex. The varied structures of the sources of the term lists are requiring modification to the data structure before terms can be imported into the master database. Database software and systems also differ.

The user interface is being developed as a Web-based entry and navigation software. Its purpose is to permit access to the widely distributed and heterogeneous source data provided by the institutions with collections and through the national nodes. Keywords will be derived from the thesaurus and also from newly developed or applied tools that generate them automatically. The mechanisms will be implemented at the central node to allow access to specimen- and observation-level data. Input by users will be possible as free text or as keywords, the latter being selected from a network of interrelated terms.

Intellectual property issues are addressed by a common code of conduct (BioCASE 2004) to which users, portals and providers in the BioCASE network adhere. In addition, ABCD offers extensive opportunities for providers to express IPR and other rights. The code of conduct ensures explicitly that such provisions made by the providers are handed on with the information to any third party wanting to use or provide the data.

4.5 EXPANDING, ENHANCING AND SUSTAINING THE NETWORK: ENBI AND SYNTHESYS

Details of ENBI (the European Network for Biodiversity Information), with 66 members in 24 countries, are given by Los and Hof in Chapter 2. The relationship of ENBI to the data access projects lies in its core objective of cementing the strong European database network that has been emerging over many years and exploring ways to sustain it. In practical terms, among many other tasks, it will increase further the number of specimen databases linked to the BioCASE system. In its capacity as representing the European contribution to the GBIF, ENBI has an implicit sustaining role for the achievements of BioCISE, ENHSIN and BioCASE.

ENBI is a thematic network. As with BioCASE and ENHSIN, it is established within the EU's Framework Programme 5. It is encouraging that the European Commission has recognized, through its funding initiatives, the value of providing transnational access to what Malacarne (2002) has described as one of 'a selected group of outstanding research infrastructures'.

Interaction between specialist users within European institutions has been a way of life since the time of Linnaeus, when travel remained difficult and computers unimagined. But interactions alone do not create infrastructures, and it is only recently that there has been a determined attempt to build an integrated system. Further success for collections infrastructures has come about through the SYNTHESYS project [5], a consortium of 19 European natural history museums and botanic gardens.

SYNTHESYS is funded as an Integrated Infrastructure Initiative within the EU's Framework Programme 6 and led by the Natural History Museum, London. One of the two major components of the project is to enable researchers to gain physical access to earth- and life science collections, facilities and expertise at 19 European institutions. The second, termed 'networking activities', is intended to create a virtual museum service and introduce further innovations to the distributed network of European collection databases.

Resources will enable the continuation of BioCASE activities until the end of 2008, well beyond the formal conclusion of that project in January 2005. Among these are a help-desk function for data providers to assist in the installation and maintenance of database wrappers; tools to improve databases structurally and with respect to data standardization and quality; means to identify duplicate specimens on the network as a means to speed up data entry; further development of the modular user interface; standardization issues; and techniques for data quality assessment and improvement.

4.6   CONCLUSION

Although creating a network of biological collection databases requires, and is receiving, attention from the international community beyond Europe, there are three main reasons why a strong European focus on this project has emerged. First, the European Commission is materially supporting the development of natural science collections as an infrastructure. Second (and closely associated with the first point) is the establishment of the European Research Area, which is predicated on the need for better integration of research and its facilities within the region. Wider cooperation exists in intercontinental forums (e.g., the Taxonomic Database Working Group, TDWG) and projects, but European expertise plays a significant role in global progress. Third, the wide spatial and temporal coverage of collections in European institutions means that they share common problems and need common solutions.

Care should be taken to ensure that providing access to data in natural science collections does not get divorced from the function of these collections. Wheeler, Raven and Wilson (2004) make the point that it is naive to '…see the information technology challenge [for taxonomy] as liberating data from cabinets'. They note rightly that providing access to bad data is unacceptable. While we do not dissent from this view, we believe that building a high-quality, virtual infrastructure of natural science collections is a task that will help expedite the very revisions for which these authors and, we suggest, the entire collections community, wish.

That there is an awareness of the problem is well illustrated by Bailly (2003; see earlier comments). There is no quick means of improving the data access infrastructures, but some progress is being made with automating data cleansing. A period of 100 years elapsed between the time of Linnaeus and the middle of the nineteenth century, when there was an explosive increase in number of species of organisms described (e.g., Gaston et al., 1995, for moths of the Lepidoptera family Geometridae). We are at an early, and probably rather crude, stage in the process of developing electronic access to data in natural science collections; mobilizing data is the current priority. Certainly, problems of data quality need thought and are there to be addressed. Yet it would surely be wise to make a concerted effort to improve access to the information of what is indeed an uneven sample, but the only truly long-term sample that we have. Taxonomy is founded on specimens.

REFERENCES

ABCD (2002) Working Group on Access to Biological Collection Data (www.bgbm.org/TDWG/ CODATA/).

Bailly, N. (2003) Functionality for satisfying user demand. In *ENHSIN: The European Natural History Information Network*, ed. M.J. Scoble. Natural History Museum, London, 133–148. (http://www.nhm.ac.uk/science/rco/enhsin/publication/ENHSIN_ch_08.pdf).

Berendsohn, W.G., ed. (2000) *Resource identification for a biological collection information service in Europa*. Botanic Garden and Botanical Museum, Berlin–Dahlem, iv, 76 pp.

Berendsohn W.G. (2003) ENHSIN in the context of the evolving global biological collections information system. In *ENHSIN: The European Natural History Information Network*, ed. M.J. Scoble. Natural History Museum, London, 21–32 (http://www.nhm.ac.uk/science/ rco/enhsin/publication/ENHSIN_ch_02.pdf).

Berendsohn, W.G., Anagnostopoulos, A., Hagendorn, G., Jakupovic, J., Nimis, P.L., Pankhurst, R.J., and White, R.J. (1999) A comprehensive reference model for biological collections and surveys. *Taxon* 48: 511–562.

Berendsohn, W.G., Costello, M.J., Emblow, C., Güntsch, A., Hahn, A., Koenemann, J., Thomas, C., Thomson, N., and White, R. (2000) Concepts for a European portal to biological collections. In *Resource identification for a biological collection information service in Europa*, ed. W.G. Berendsohn. Botanic Garden and Botanical Museum, Berlin–Dahlem, 59–71.

BioCASE (2004) BioCASe code of conduct for data nodes (data providers) and portal nodes for sharing of unit data (http://www.biocase.org/Doc/Doc/BioCASE-Code-of-Conduct.shtml).

Copp, C.J.T. (2003) Creating and managing a thesaurus for accessing natural science collection and observation data. In *ENHSIN: The European Natural History Information Network*, ed. M.J. Scoble. Natural History Museum, London, 149–164 (http://www.nhm.ac.uk/science/rco/ enhsin/publication/ENHSIN_ch_09.pdf).

Gaston, K.J., Scoble, M.J., and Crook, A. (1995) Patterns in species description: A case study using the Geometridae. *Biological Journal of the Linnean Society* 55: 225–237.

GBIF (2003) GBIF Work Programme 2004. Approved by the GBIF Governing Board at GB7 October 2003, Tsukuba, Japan (http://www.gbif.org/GBIF_org/wp/wp2004).

Güntsch, A. (2003) The ENHSIN pilot network. In *ENHSIN: The European Natural History Information Network*, ed. M.J. Scoble. Natural History Museum, 33–40 (http://www.nhm.ac.uk/ science/rco/enhsin/publication/ENHSIN_ch_03.pdf).

Hahn, A. (2000) Information resources — The BioCISE Survey. In *Resource identification for a biological collection information service in Europa*, ed. W.G. Berendsohn. Botanic Garden and Botanical Museum, Berlin–Dahlem, 39–48.

Malacarne, M. (2002) Foreword. In *The European Natural History Specimen Information Network*, ed. M.J. Scoble. Project report, European Commission.

Wheeler, Q.D., Raven, P.H., and Wilson, E.O. (2004) Taxonomy: Impediment or expedient? *Science* 303: 285.

CITED WWW RESOURCES

1. BioCISE: http://www.bgbm.fu-berlin.de/biocise/default.htm
2. ENHSIN: http://www.nhm.ac.uk/science/rco/enhsin/
3. BioCASE: http://www.biocase.org
4. ENBI: http://www.enbi.info
5. SYNTHESYS: http://www.synthesys.info
6. 'Species Analyst': http://speciesanalyst.net/
7. REMIB: http://www.conabio.gob.mx/remib_ingles/doctos/remib_ing.html
8. ENHSIN 'element set': http://www.bgbm.fu-berlin.de/BioDivInf/projects/ENHSIN/Pilot Implementation.htm
9. BioCASE Project: http://www.biocase.org/Doc/Project/DoW/DoW-Summary.shtml

5 *A Comparison* between Morphometric and Artificial Neural Network Approaches to the Automated Species Recognition Problem in Systematics

Norman MacLeod, M. O'Neill and Steven A. Walsh

CONTENTS

ABSTRACT

One approach to addressing long-standing concerns associated with the taxonomic impediment and occasional low reproducibility of taxonomic data is through development of automated species identification systems. Such systems can, in principle, be combined with image-based or image- and text-based taxonomic databases to add elements of expert system functionality. Two generalized approaches are considered relevant in this context: morphometric systems based on some form of linear discriminant analysis (LDA) and artificial neural networks (ANNs). In this investigation, digital images of 202 specimens representing seven modern planktonic foraminiferal species were used to compare and contrast these approaches in terms of system accuracy, generality, speed and scalability. Results demonstrate that both approaches could yield systems whose models of morphological variation are over 90% accurate for small data sets. Performance of distance- and landmark-based LDA systems was enhanced substantially through application of least-squares superposition methods that normalize such data for variations in size and (in the case of landmark data) two-dimensional orientation. Nevertheless, this approach is practically limited to the detailed analysis of small numbers of species by a variety of factors, including the complexity of basis morphologies, speed and sample dependencies. An ANN variant based on the concept of a plastic self-organizing map combined with an *n*-tuple classifier was found to be marginally less accurate, but far more flexible, much faster and more robust to sample dependencies. Both approaches are considered valid within their own analytic domains, and both can be usefully synthesized to compensate for their complementary deficiencies. Based on these results (as well as others reviewed here), it is concluded that fast and efficient automated species recognition systems can be constructed using available hardware and software technology. These systems would be sufficiently accurate to be of great practical value notwithstanding the fact that the already impressive performance of current systems can be improved further with additional development.

5.1   INTRODUCTION

5.1.1   THE NEED FOR AUTOMATED SPECIES RECOGNITION IN SYSTEMATICS

The automated identification of biological species has been something of a holy grail among taxonomists and morphometricians for several decades. Many multivariate morphometrics textbooks of the 1970s and 1980s contained chapters dealing with aspects of the discrimination problem, often basing those discussions on R.A. Fisher's classic treatment of discriminations among three *Iris* species (e.g., Sokal and Sneath 1963; Blackith and Reyment 1971; Pimentel 1979; Neff and Marcus 1980; Reyment et al. 1984). Despite these introductions to the quantitative side of the object classification problem, progress in designing and implementing practical systems for fully automated species identification has proven frustratingly slow. Discounting passive taxonomic databases, some of which contain semi-automated interactive keys (e.g., MacLeod 2000, 2003), we are aware of no such systems in routine operation within any area of biological or palaeontological systematics.

The reasons for this lack of progress are many-fold. Development of such systems presents a formidable challenge that, until recently, was beyond the technological capabilities of existing information technology. Even though these hardware limitations of such systems have largely been addressed, software development remains complex and well beyond the programming skills of most classically trained systematists. This, combined with (1) a lack of interest in and appreciation of the subtleties of taxonomic identification by most programming specialists, mathematicians, artificial intelligence experts, etc.; (2) the enormous range of morphologies that must be dealt with in order to construct a practical identification system for any but trivial purposes; and (3) a genuine reticence on the part of the systematics community to prioritize such a technology-driven research programmes have (we believe) conspired to limit the progress that clearly needs to be achieved in this area.

The reasons why progress in this area must be made are also manifold. Perhaps most important of these is the looming taxonomic impediment. Put crudely, the world is running out of specialists who *can* identify the very biodiversity whose preservation has become such a global concern (e.g., Gaston and May 1992). This expertise deficiency cuts as deeply into those commercial industries that rely on accurate species identifications (e.g., agriculture, biostratigraphy) as it does into the capabilities of a wide range of pure and applied research programmes (e.g., conservation, biological oceanography, climatology, ecology). While most scientists recognize the existence and serious implications of this phenomenon, hard data on the taxonomic impediment's size are difficult to come by.

One indication, however, is provided by a recent American Geological Institute report on the status of academic geoscience departments that shows that, between the 1980s and 1990s, the number of palaeontology–stratigraphy theses and dissertations completed per annum declined by 50%, and the number of palaeontology–stratigraphy faculty positions fell by a greater amount than for any other geoscience discipline (e.g., geophysics, structure/ tectonics). Moreover, the average age of geoscience faculty members in 2000 was almost twice the average age in 1986. In commenting on this problem in palaeontology as long ago as 1993, Roger Kaesler recognized the following:

> …[W]e are running out of systematic paleontologists who have anything approaching synoptic knowledge of a major group of organisms [p. 329]. Paleontologists of the next century are unlikely to have the luxury of dealing at length with taxonomic problems…[and] will have to sustain its level of excitement without the aid of systematists, who have contributed so much to its success [p. 330].

A second reason why research effort is needed in the systematic application of automated object recognition technology centers around the need to improve the consistency and reproducibility of taxonomic data. At present it is commonly, though informally, acknowledged that the technical, taxonomic literature of all organismal groups is littered with examples of inconsistent and incorrect identifications (e.g., Godfrey 2002). This is due to a variety of factors, including authors being insufficiently skilled in making distinctions between species; insufficiently detailed original species descriptions and/or illustrations; authors using different rules of thumb in recognizing the boundaries between morphologically similar species; authors having inadequate access to the current monographs and well-curated collections; and, of course, authors having different opinions regarding the status of different species concepts. Peer review only weeds out the most obvious errors of commission or omission in this area and then only when the author provides adequate illustrations of the specimens in question. Systematics is not alone among intellectual disciplines in confronting problems of this sort, but systematics is well behind other sciences in making progress toward their resolution or, indeed, even in acknowledging their scope.

Another reason for considering an automated approach to the species identification problem is that classical systematics has much to gain, practically and theoretically, from such an initiative. It is now widely recognized that the days of taxonomy as the individualistic pursuit of knowledge about species in splendid isolation from funding priorities and economic imperatives are rapidly drawing to a close. In order to attract personnel and resources, morphology-based taxonomy must transform itself into a 'large, coordinated, international scientific enterprise' (Wheeler, 2003, p. 4). Many have recently touted use of the Internet, especially via the World Wide Web, as the medium through which this transformation can be made (e.g., Godfrey 2002; Wheeler 2003; Wheeler et al. 2004). While establishment of a virtual, GeneBank-like system for accessing morphological information would be a significant step in the right direction (see MacLeod 2002a), improved access to specimen images and text-based descriptions alone will not address the taxonomic impediment or low reproducibility issues successfully.

Instead, the inevitable subjectivity associated with making critical decisions on the basis of qualitative criteria must be reduced or, at the very least, embedded within a more formally analytical context. A properly designed, flexible, robust, automated species recognition system organized around the principles of a distributed computing architecture can, in principle, produce such a system.

In addition, the process of taxonomic identification must be endowed with better ways of capturing the memory and preserving the reasoning behind particular taxonomic decisions so that these can be reconstructed objectively and independently for subsequent evaluation. This would allow taxonomy to accumulate information over time in a much more efficient way than it does now and so achieve the highly desirable property of ever increasing accuracy through use. Continued reliance on individualistic and entirely qualitative forms of identification and data recording will not achieve this goal.

To be of optimal use, an automated identification system could be designed to operate in authoritative (for routine identifications) or interactive modes, the latter of which could be used by specialists to develop and/or test hypotheses of character-state identification/distribution that bear on the question of species discrimination and/or group membership. In this way, such systems could function as active partners in systematic research as well as passive bookkeepers or databases of research results, even to the point of checking existing museum collections for identification correctness and consistency. Finally, all this must be done in a manner that does not impose particular types of species concepts on users or constrain the types of information that can be used to delineate taxonomic groups.

5.1.2 APPROACHES

To date, there have been two generalized approaches to the design of systematic species recognition systems. The morphometric approach (Figure 5.1A) uses a series of linear distance variables or landmarks to quantify the size and size/spatial distribution (respectively) of a specimen's morphological features relative to one another (e.g., Young et al. 1996). By sampling aspects of the morphology that characterizes known species in the form of training sets of authoritatively identified specimens, models of intraspecific variation can be constructed. Models so constructed for different species can then be contrasted with one another using a variety of multivariate procedures (e.g., cluster analysis, principal components analysis, discriminant analysis, canonical variates analysis).

These methods use the selected aspects of the specimen's size and shape to construct a continuous, multivariate feature space within which all members of the training set may be located. Once constructed this biologically determined (by virtue of the measurements selected) feature space can be used to define partitions within this space that delimit the boundaries between the *a priori* training set groups. Unknown specimens can then be identified by collecting these same data, using them to project the specimen into the partitioned feature space, and assigning it to the group into whose partition it falls. (Note: Depending

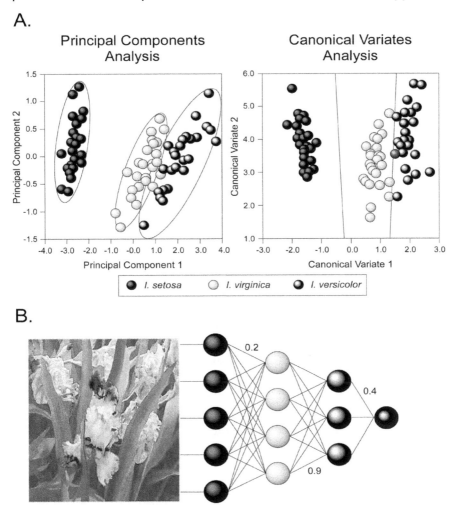

FIGURE 5.1   Alternative conceptual approaches to the species identification problem. A. Linear multivariate approaches use covariance or correlation indices to assess the structure of biologically meaningful geometric relations between individuals (e.g., principal components analysis) or between groups (e.g., canonical variates analysis) and then employs these to construct an optimized linear, multidimensional, feature space that can be subdivided into group-specific domains. B. Artificial neural networks use layers of switches that can be assigned variable weights connected into a network. These switch arrays can then be trained to discriminate between objects based on generalized input data fed into each switch through recursive, trial and error weight adjustment. Once the network has been trained, the weight scheme can, in principle, be used to construct a generalized, non-linear, multidimensional feature space.

on how the intergroup partitions are defined, the object may fall outside the range of any species whose limits have been established by this method, in which case the object would remain unassigned.)

    The second approach to automated object recognition uses a computational approximation of human neural systems — an artificial neural net, or ANN — to achieve discrimination (Figure 5.1B). The 'neurons' of this system are switches designed to open or remain closed based on the strength of generalized input signals (e.g., pixel brightness values). Banks of these artificial neurons are arranged in two or more series; the connections between neurons are able to be assigned numerical weights that amplify or diminish the strength of the signal as it passes along interneuron paths (Bishop 1995; Ripley 1996; Schalkoff 1997).

Instead of partitioning a selected measurement-defined feature space, ANNs achieve discrimination by being trained on inputs from *a priori* training sets of authoritatively identified specimens. This training amounts to recursive adjustment of the interneuron weights until the desired output (optimal identification of training set objects) is achieved. Once an optimum weight scheme has been determined on the basis of these training sets, unknown objects are identified by submitting their input signals to the system. Because of the more general nature of the ANN switches and the fact that the weight scheme is determined recursively, ANN systems utilize a greater variety of input observations than morphometric approaches.

Both approaches have advantages and disadvantages. Morphometric systems are potentially more efficient for well-defined data sets of similar morphologies because they can concentrate on morphological features known or suspected to be reliable species discriminators. Morphometric systems can, however, also become limited if the best morphological targets for group discrimination are unknown, if the morphology is sufficiently complex (so as to render automated feature extraction and/or measurement from images unreliable) or if the morphology is sufficiently simple (so as to reduce the number of common and consistently expressed morphological features available for measurement). Artificial neural networks can accommodate a greater variety of input signals (e.g., pixel brightness and/or colour values), but the ability to work with greater amounts and more generalized types of spatial information can make signal extraction more difficult. Standard, or supervised, ANNs can suffer from being time consuming to tune. Bollmann et al. (2004, p. 14) noted that tuning of the COGNIS supervised ANN system on image set of 14 coccolith species containing 1000 images took 'several hours', while tuning for a two-species 2000-image set took 'over 30 hours'.

Both morphometric and supervised ANN approaches also suffer from the fact that their weight schemes are linked deterministically to the group-level contrasts over which they have been optimized. Consequently, addition of even a single new species to the set requires complete recalibration of all multivariate feature space partitions and weight schemes for the interneuron connections. Finally, there is the practical issue of scalability. In order to be practical, an automated object recognition system must be able to extract unique features from and be optimized over hundreds of species whose morphological distinctions range from the obvious to the very subtle.

One recent development in ANN technology that addresses some deficiencies of supervised ANNs has been the development of unsupervised variants such as Kohonen-based algorithms, including plastic self-organizing maps (PSOM; Lang and Warwick 2002), which are variants of Lucas continuous n-tuple classifiers (Lucas 1997). This type of ANN incorporates an aspect of artificial intelligence (dynamic learning) into its algorithms that obviates the need to recalibrate the interneuron weight scheme completely. Under this approach, such recalibrations as are necessary can usually be handled in real time as new training sets are added to the system. Gaston and O'Neill (in press) report that n-tuple/ PSOM systems also respond well to the modeling of non-linear regions within shape–space distributions, which are known to be problematic for many (though not all) types of morphometric approaches (Bookstein 1991).

5.1.3 OBJECTIVES

Owing to the importance of achieving a robust solution to the automated object recognition problem in biological taxonomy and to the potential of recent developments in the area of unsupervised ANN technology, we intend to begin a systematic evaluation of the various approaches to this generalized problem here, with a comparison of relative levels of performance between distance- and landmark-based canonical variates analysis (currently the most popular morphometric method for achieving group-based discriminations) and an implementation of the *n*-tuple/PSOM approach (the most advanced of the ANN-based techniques, but one that has yet to be tested directly against any alternative method). The objectives of this investigation are fourfold: to compare and contrast the (1) accuracy; (2) generality; (3) speed; and (4) scalability of these approaches. This comparison will focus entirely on species recognition aspects of the system design problem; no effort will be devoted to addressing the issues of automated image acquisition or automated feature extraction (see Bollmann et al. 2004).

The subjects of this test will be a set of images of seven modern planktonic foraminiferal species picked from core-top sediments collected from the western Atlantic Ocean. Planktonic foraminifera represent very desirable subjects for this type of investigation because

- their systematics is based entirely on morphological features;
- they are studied and identified entirely through the use of two-dimensional, remote images;
- their taxonomy is stable and well known;
- they are used in a wide variety of scientific contexts (e.g., oceanography, biogeography, marine ecology, climatology);
- a small number of species can encompass a large proportion of the total morphological diversity; and
- they constitute a morphologically representative subset of a large, but not enormous, fossil fauna that has considerable utility in an even broader array of contexts (e.g., foraminiferal systematics is a key biostratigraphic tool for petroleum exploration).

In other words, success in constructing a practical and reliable system for automatically identifying planktonic foraminiferal species should have considerable economic as well as intellectual and symbolic value.

5.1.4 MATERIALS AND METHODS

This comparison was conducted on a small sample of monochrome digital images of seven planktonic foraminiferal species (Figure 5.2). Representative specimens of each species were picked randomly from a Vema Cruise core-top sample (sample no. V24-99 50) collected from the Baltimore Canyon, offshore New Jersey, USA. All images were taken with a colour digital video camera at relatively low resolution (72 dpi). Aside from photographing all specimens in umbilical view, no extraordinary attempts were made to correct specimen orientation or use composite images to improve image quality. The reason for this was that, in order to be practical, any automated species identification system will need to work with images that can be collected quickly, inexpensively and in as automated a manner as possible. Likewise, all images were brought to a consistent exposure using the autolevel

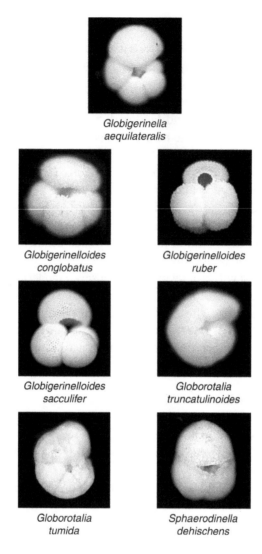

*Globigerinella
aequilateralis*

*Globigerinelloides
conglobatus*

*Globigerinelloides
ruber*

*Globigerinelloides
sacculifer*

*Globorotalia
truncatulinoides*

*Globorotalia
tumida*

*Sphaerodinella
dehischens*

FIGURE 5.2 Planktonic foraminiferal species used in this investigation with representative illustrations of image qualities used to assess two-dimensional patterns of intraspecific variation. These images were captured quickly, using standard resolution video cameras with no time taken for fine adjustment of exposure, depth of field or specimen orientation.

routines of standard image processing software (e.g., Adobe Photoshop, Graphic Converter) running in scripted mode.

For morphometric analysis, coordinate data for a set of 11 discrete landmarks were collected from each specimen's image (Figure 5.3). Because of limited morphological homology among these species in umbilical view geometric data could only be collected from the final three chambers and approximated the coordinate positions of each chamber's major axes. In principle, these data could have been taken from each specimen without having to capture the specimen's image. In order to ensure comparability with the ANN results, however, all landmark coordinates were collected from the same images employed in the ANN analysis. In order to evaluate the best type of morphometric data for use in this context,

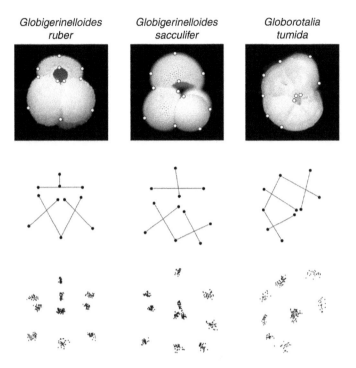

FIGURE 5.3   Morphometric data types used in this investigation. Each specimen (upper row) was characterized morphologically through measurement of the coordinate locations of 11 landmarks that quantify the major dimensions of the last three chambers (ultimate, penultimate and prepenultimate). These landmarks were then used to construct data sets of interlandmark distances (middle row) and superposed landmark arrays (bottom row).

these landmark points were used to represent morphological variation as a set of six interlandmark distances (the classical morphometric variables) and as raw x,y coordinate locations (the preferred geometric morphometric variable type).

Two sets of distance data were constructed, one from the raw landmark coordinates and the other from the coordinate locations after least-squares superposition (Bookstein 1991). This allowed evaluations of size-referenced and size-normalized representations of morphological variation to be evaluated for their interspecific discriminant power. In the case of the purely landmark-based analysis, only superposed landmarks were used, as is typical of geometric morphometric analyses.

Multivariate discriminant analysis was carried out on these data using canonical variates analysis (CVA; see Blackith and Reyment 1971; Pimentel 1979; Reyment et al. 1984). Each training set was constructed from measurements (see earlier discussion) taken from the images of authoritatively identified specimens. No additional data transformations were carried out prior to CVA analysis.

As noted by Campbell and Atchley (1981), CVA performs within-group, variance–covariance standardization prior to between-groups eigenanalysis. When applied to superposed landmark data directly, this has the effect of distorting the Procrustes distance metric for representing within-group relations among specimens. Because of this standardization, use of CVA and related approaches (e.g., MANOVA, MANCOVA) should always be applied with caution to such data. Specifically, no attempts should be made to interpret the details

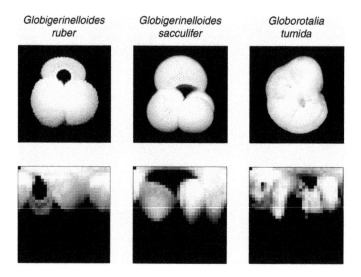

FIGURE 5.4 Examples of input for the artificial neural network trial. Each specimen's image (upper row) was subsampled to a 32 × 32 pixel grid, standardized for variations in exposure using image-histogram equalization, and transformed from a Cartesian to a polar pixel coordinate system (bottom row). The RGB brightness values for each pixel constitute a multivariate vector representing each image. These values correspond to the measurements and landmark coordinates used as observations in the morphometric data analyses.

of within-group ordinations within the shape spaces defined by CVA axes. The geometry of between-groups ordinations are more faithfully preserved in such spaces, but even these may be distorted relative to results obtained by methods specifically designed to preserve the landmark-based Procrustes metric (e.g., relative warps analysis, coordinate-point eigenshape analysis). Throughout, it must be kept in mind that the appropriate use of such methods is restricted to testing the hypothesis of *a priori* group distinctiveness in a multivariate context and facilitating the identification of objects based on measurement sets that can be projected into the (distorted) canonical variates shape space.

The PSOM/*n*-tuple ANN approach to species identification was implemented by the digital automated image-analysis system (DAISY; Weeks et al. 1997, 1999a, b). This implementation accepts training sets in the form of standard format images (e.g., jpeg, tiff) of authoritatively identified specimens. These image-based training sets were processed (1) by reducing each image's spatial resolution (via subsampling) to a 32 × 32 pixel grid; (2) by transforming each image's 32 × 32 pixel grid from a Cartesian to a polar format (Figure 5.4), and 3) by adjusting each image's pixel-level spectrum to achieve brightness histogram equalization. The first step in this process represents an empirically determined optimum resolution needed to maximize the signal-to-noise ratio and quantify topological correspondences. The second allows the analysis to utilize spatially irregular regions of interest as well as the more traditional rectilinear image boundaries. The third reduces interimage variations and renders the image input easy to correct for the effects of inconsistent pose due to lighting/exposure artefacts.

Once DAISY had processed all images in the training set, a discriminant space was calculated by applying the PSOM/*n*-tuple classifier to the training set composed of the polar-formatted, 32 × 32 pixel images. The proximate basis for this classification is a pairwise comparison between brightness values between pixel locations. The result allows each object in each training set to be placed into a multidimensional, distance-based ordination space whose character can be varied based on the estimated affinity (estimated via cross-correlation) between similarly processed images of unknown specimens and the training-set array. It is this ability to modify the character of the base training set ordination that gives the DAISY implementation of the unsupervised ANN concept its adaptive quality. Identifications are achieved by assessing the eightfold nearest-neighbour coordination between each unknown and the training-set ordination.

5.1.5   RESULTS

Table 5.1 summarizes the cross-validation results for each of the four analyses. Each analysis returned results that were highly accurate and consistent with the overwhelming majority of training set measurements being allocated to their correct groups within the empirical discrimination space. Nevertheless, each result also reveals strengths and weaknesses of the respective analytic approaches and data.

The traditional, interlandmark distance-based CVA returned 91% correct cross-validated identifications for the 202 specimens based on six generalized distances taken from the ultimate, penultimate and prepenultimate chambers in umbilical view. This result is unexpectedly high owing to the fact that neither the absolute nor the relative dimensions of these final three chambers have been judged to be critical to the correct identification of any of these species previously (e.g., Kennett and Srinivasan 1983; Bolli and Saunders 1985). Typical raw distance-based, cross-validation analyses for marine microplankton yield correctness ratios of 0.7 to 0.9 (e.g., see Culverhouse et al. 2003). This isolated correct identification score can be misleading, however, unless it is put into context by summarizing the strength of support for each identification. This is especially important in that the robust identification of unknown objects should be undertaken in light of precisely such assessments.

Examination of the posterior identification probabilities for the data set taxa (summarized in Figure 5.5) provides a more nuanced understanding of the result. Of the 202 specimens used to construct the discriminant space, 184 were identified correctly. Of these, only 114 (62%) were identified with a probability of 0.95 or higher. Taking these results, in addition to the incorrect identifications, into consideration this data set exhibits a confident identification ratio (probability ≥ 0.95) of only 0.56.

One factor affecting the discrimination efficiency of raw, interlandmark distance data is the confounding of size and shape variation. Each of these seven species exhibits a range of sizes with much between-species overlap and distinction (Figure 5.6). Yet, the primary features used for qualitative species identification are shape differences between component parts of the organism's skeleton.

Using the least-squares superposition method (Bookstein 1991), it is possible to standardize these landmark data for generalized size differences and then recalculate the interlandmark distances so that they form a more faithful summary of distinctions solely attributable to between-species shape differences. When these size-standardized distances are used to construct the discriminant space, the raw ratio of correct cross-validation identifications rises to an impressive 0.96 (Table 5.1). Even more impressive, though, are the improvements in the amount of statistical support available for these identifications (Figure 5.7). Of the 193 specimens identified correctly, 154 (80%) had a posterior correct identification

TABLE 5.1

Results of Cross-Validation Tests for Canonical Variates Analysis (CVA) and Artificial Neural Network Analysis (DAISY) of 202 Planktonic Foraminiferal Specimens

	Ge. aequilat.	Gl. conglob.	Gl. ruber	Gl. sacculifer	Gr. truncat.	Gr. tumida	S. dehiscens	Total	Correct
Raw distance-based CVA									
Ge. aequilateralis	20	0	0	1	0	5	0	26	0.77
Gl. conglobatus	0	30	1	0	0	0	0	31	0.97
Gl. ruber	0	2	37	0	0	0	0	39	0.95
Gl. sacculifer	0	0	0	33	0	0	0	33	1.00
Gr. truncatulinoides	0	0	0	1	23	0	2	26	0.88
Gr. tumida	4	0	0	0	0	20	0	24	0.83
S. dehiscens	0	0	0	0	2	0	21	23	0.91
Total correct	24	32	38	35	25	25	23	202	0.91
Superposed distance-based CVA									
Ge. aequilateralis	23	0	0	0	0	3	0	26	0.88
Gl. conglobatus	0	29	1	1	0	0	0	31	0.94
Gl. ruber	0	0	39	0	0	0	0	39	1.00
Gl. sacculifer	0	1	0	32	0	0	0	33	0.97
Gr. truncatulinoides	0	0	0	1	25	0	0	26	0.96
Gr. tumida	0	0	0	0	0	24	0	24	1.00
S. dehiscens	0	1	0	0	1	0	21	23	0.91
Total correct	23	31	40	34	26	27	21	202	0.96

Superposed landmark-based CVA

	Ge. aequilateralis	Gl. conglobatus	Gl. ruber	Gl. sacculifer	Gr. truncatulinoides	Gr. tumida	S. dehiscens		
Ge. aequilateralis	25	0	0	0	0	1	0	26	0.96
Gl. conglobatus	0	30	1	0	0	0	0	31	0.97
Gl. ruber	0	0	39	0	0	0	0	39	1.00
Gl. sacculifer	0	0	0	33	0	0	0	33	1.00
Gr. truncatulinoides	0	0	0	0	26	0	0	26	1.00
Gr. tumida	0	0	0	0	0	24	0	24	1.00
S. dehiscens	0	0	0	0	0	0	23	23	1.00
Total correct	25	30	40	33	26	25	23	202	0.99

DAISY

	Ge. aequilateralis	Gl. conglobatus	Gl. ruber	Gl. sacculifer	Gr. truncatulinoides	Gr. tumida	S. dehiscens		
Ge. aequilateralis	26	0	0	0	0	0	0	26	1.00
Gl. conglobatus	0	30	0	1	0	0	0	31	0.97
Gl. ruber	0	0	39	0	0	0	0	39	1.00
Gl. sacculifer	0	0	1	31	0	0	0	33	0.94
Gr. truncatulinoides	0	0	0	0	26	0	0	26	1.00
Gr. tumida	0	0	0	0	0	24	0	24	1.00
S. dehiscens	0	0	0	0	0	0	23	23	1.00
Total correct	26	30	40	32	26	24	23	202	0.99

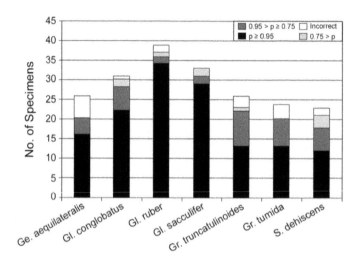

FIGURE 5.5 Histogram of posterior probabilities for the cross-validation study of the raw, interlandmark distance-based canonical variates analysis. Different shaded boxes represent numbers of specimens included in various degree of support categories. See text for discussion.

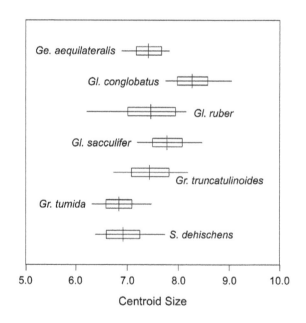

FIGURE 5.6 Size variation in the seven planktonic foraminiferal data set used in this investigation. Horizontal line indicates range of centroid-size values. Open box represents ±1.0 standard deviations from the mean. Vertical lines indicate position of the sample means. Note wide degree of size variation within and between species.

probability of 0.95 or higher. Thus, simply standardizing interlandmark distance data for size variation resulted in an increase in the number of confident identifications by 24%.

Of course, for the past 15 years the field of morphometrics has been moving away from the use of interlandmark distance measurements in favour of statistical operations on the two- or three-dimensional landmark coordinates (e.g., Bookstein 1986, 1991; Rohlf and

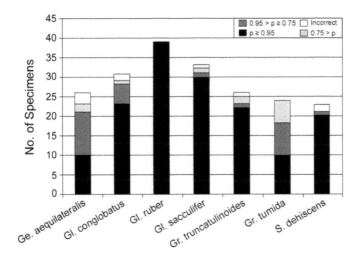

FIGURE 5.7   Histogram of posterior probabilities for the cross-validation study of the superposed, interlandmark distance-based canonical variates analysis. Differently shaded boxes represents numbers of specimens included in various degree of support categories. See text for discussion.

Bookstein 1990; Marcus et al. 1993, 1996; MacLeod and Forey 2002). These variables have the advantage of being able to quantify a much greater proportion of the underlying morphology than can be assessed with scalar distances alone. In terms of the present analysis, use of the 11 landmark coordinates captures aspects of chamber size, chamber shape, chamber orientation, relative degree of chamber inflation, chamber appression, the number of chambers in the final whorl, height of the primary aperture, degree of interchamber suture incision, umbilicus position, umbilicus size and umbilicus shape. Unlike the directed scalar distances used in the first two analyses, many of these characters are considered important in the specific diagnosis of these species (see Kennett and Srinivasan 1983; Bolli and Saunders 1985).

Once again, using least-squares superposition to normalize the coordinate data for generalized size differences (thereby achieving an entirely shape-based discrimination) and employing CVA to construct a discriminant space, an unprecedented correct cross-validation identification ratio of 0.99 was obtained (Table 5.1 and Figure 5.8). Of the two misidentified specimens, a *Globigerinelloides conglobatus* was mistaken completely for *Globigerinelloides ruber* (posterior probability = 1.00) while a *Globigerinella inaequilateralis* was ambiguously mistaken for *Globorotalia tumida* (posterior probability = 0.67).

Cross-validation results for the DAISY-based ANN analysis differ from those of the CVA analysis in terms of the manner in which the posterior probabilities are calculated. Instead of using a distance-based approach for assigning unknowns to groups, DAISY uses a combined eightfold distance-coordination approach with the minimum coordination value for identification set to three. This amounts to projecting each unknown into a discrete feature space and determining the identity of its eight nearest neighbors. Once these identities are known, a variety of statistical measures of the strength of support for a particular identification can be generated.

However, because only eight known comparators are used to evaluate the support strength of each identification, the posterior probability scale is discrete rather than continuous and falls off rapidly if there is any disagreement in group membership. For example, if the

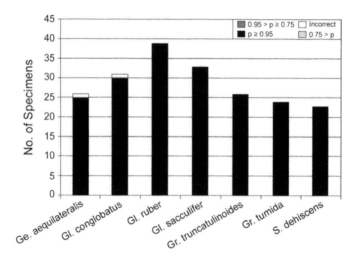

FIGURE 5.8 Histogram of posterior probabilities for the cross-validation study of the superposed, landmark coordinate-based canonical variates analysis. Differently colored boxes represent numbers of specimens included in various degree of support categories. See text for discussion.

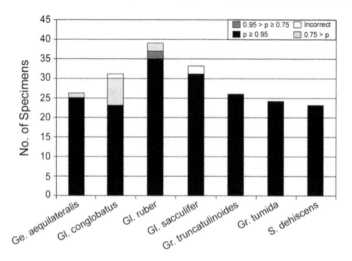

FIGURE 5.9 Histogram of posterior probabilities for the cross-validation study of the DAISY-based PSOM/n-tuple artificial neural network analysis. Differently shaded boxes represent numbers of specimens included in various degree of support categories. See text for discussion.

images of an unknown specimen's six nearest neighbors all belong to group 1 and those of the two remaining nearest neighbors belong to group 2, the strength of support is reported as 0.75. This biases the DAISY results against high posterior probability values for any identification that is less than perfect, but it also results in the imposition of a very conservative rule base for making identification decisions.

Despite the far more generalized nature of the data used to construct the feature space and the less unforgiving rules used for determining identifications, the DAISY cross-validation results are fully comparable to best results that were able to be obtained through CVA (see Table 5.1), with only marginally lower posterior probabilities of identification support (Figure 5.9). In this context, it is important to note how much better DAISY performance was over performances of traditional distance-based CVA using raw or processed (super-posed) data, both in terms of raw numbers of correct identifications (0.91 vs. 0.96 vs. 0.99) and in terms of the number of well-supported (p ≥ 0.95) identifications (0.56 vs. 0.76 vs. 0.93). The only linear discriminant method that produced results comparable to those of the DAISY-based ANN implementation was a superposed landmark-based canonical variates analysis.

5.1.6 Discussion

Figure 5.10 illustrates a comparison of the results obtained by this study with those of other semi-automated and automated systems for species identification based on morphological characteristics. This comparison confirms that results obtained from superposed distance and superposed landmark CVA, along with the DAISY results for this selection of plank-tonic foraminiferal species, are among the best that have been obtained to date for compa-rably sized data sets. The obvious questions are

1. Which approach (morphometric or ANN) holds the greater promise for use in cre-ating a practical, general purpose, fully automated object recognition system?
2. Is there any scope for combining these approaches to achieve even greater perfor-mance levels?
3. What research remains to be done before such a system can be realized?
4. What should be the systematics community's attitude to these technological developments?

5.1.6.1 Which Approach?

Although superposed distance and superposed landmark LDA approaches achieved mar-ginally superior performance in terms of per cent correct identifications, there are several practical considerations that, we believe, will limit the ability of these methods to contrib-ute to solutions of the overall automated species identification problem. The foraminiferal analysis undertaken here involved a small number of species. Indeed, LDA for the purpose of species identification almost always involves a small number of species (e.g., Gaston and O'Neill, 2004). The reasons for this are twofold. First and most superficially, since such studies are not typically regarded as mainstream systematics, they tend to be — like this study — demonstrations designed to describe and explore new approaches to LDA analysis. Such demonstrations do not require large data sets because their purpose does not usually include any examination of the scalability problem.

The fact that this latter part of a more generalized challenge is rarely addressed leads to the second, more substantive difficulty. The information input necessary for application of LDA methods to medium-scale (50–100 species) and large-scale (100+ species) data sets will be practical only for very complex morphologies. As a minimum condition, any system containing n groups can only be resolved completely in a discriminant space containing $n - 1$ dimensions. Thus, the LDA solution of a 50-group problem implies the collection of 49 different variables on which to base the construction of a fully resolved LDA space. If one were to adopt a superposed landmark-based approach, this could be achieved via the specification of 25 landmarks that could be located on all taxa. However, the minimum

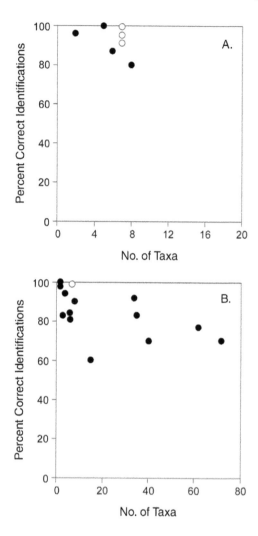

FIGURE 5.10 Comparison between the results obtained by this investigation (open circles) and those tabulated by Gaston and O'Neill (2004) for the fidelity of linear discriminant analysis (A) and artificial neural networks (B) used for automated species identification in a variety of organismal groups.

number of landmarks that can be used to describe individuals within such a measurement system is determined by the *least* morphologically complex taxon.

The operation of this principle is well illustrated by the foraminiferal analysis undertaken in this study. Even though a majority of species contain more than three chambers in their final whorl (see Figure 5.1), assessment of shape variation based on the ultimate, penultimate and prepenultimate chambers was necessitated because these were the only chambers visible in umbilical view for some of the included species (*Globigerinelloides ruber*, *Globigerinelloides sacculifer*, *Sphaerodinella dehiscens*). If, for example, the common modern planktonic foraminiferal species *Orbulina universa* had been included in the study group (see Figure 5.11), a substantial change to the measurement strategy would have been required because the adult skeleton of this species is composed of a single chamber that envelops all others, rendering the penultimate and prepenultimate chambers invisible.

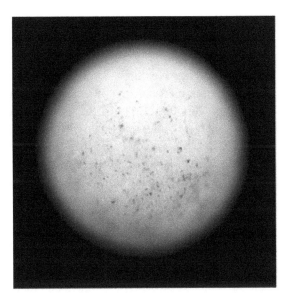

FIGURE 5.11 Example image for the planktonic foraminiferal species *Orbulina universa*. The spherical, ultimate chamber of this species completely envelops all previous chambers, hiding them from view. If this species had been included in the data set, only morphometric data from the ultimate chamber of each species would have been able to have been collected, and even then the detailed topological correspondence between landmarks collected from different species would not have been able to have been preserved. As a result, the ability of all investigated morphometric approaches to species discrimination would have been compromised severely. However, inclusion of this species would not have affected any aspect of data collection for the DAISY-based PSOM/*n*-tuple approach nor engendered any pronounced effect on its results.

The effect of basing a morphometric LDA on only the ultimate chamber shape of each species would have been to degrade the power of this analysis class severely. Under such a strict measurement protocol, it is questionable whether sufficient morphological resolution could be achieved to completely resolve the discriminant space for even the seven group problem.

It is also important to note that a necessarily corollary to the characterization of morphological variation through morphometric methods is the often time-consuming and skilled nature of the data-collection task. Even using sophisticated landmark collection software (e.g., ImageJ, tpsDig), assembly of landmark data for all 202 specimens took approximately seven hours of quite tedious work and required the technician to possess a detailed familiarity with the morphological character of each species. It is doubtful that accurate data of this type could be collected by anyone not already familiar enough with the taxonomy of the group to provide a reliable identification in much less time. [Note: While it is true that automated landmark location software does exist, these programmes must themselves be tuned to operate efficiently on different morphologies, and then tested in a similarly time-consuming, and group-limited manner.]

The DAISY implementation of the ANN approach circumvents this data collection problem by assuming that comparisons between objects useful in addressing the discrimination problem can be made on the basis of pixel matching across the entire 32 × 32 pixel frame. This approach relaxes the morphometric requirement for landmarks to represent a comparatively small number but biologically well-known set of close topological correspondences between objects in favour of more inclusive information drawn from the spatial distribution of non-specific group features. Though not as biologically sensitive as the strict morphometric data collection protocol, the DAISY/ANN approach has the desirable property of collecting a large amount of data — including some proportion of three-dimensional data — and being able to be automated completely. In our foraminiferal study, the subsampling required to match images across the data set took less than four minutes by an algorithm that was not designed to operate at maximum speed.

Such considerations lead to a series of recommendations regarding the future roles of these two generalized approaches to automated object recognition. First, morphometric and ANN approaches should not be seen as competitors, but rather as complements, each with marked strengths within its own domain. The domain of morphometric analysis is that of investigating biologically meaningful comparisons between forms. This biological meaning is provided by the selection of landmarks. Thus, the feature space within which morphometric comparisons are made is explicitly biological and incontrovertibly tied to the analyst's mapping of biological meaning onto the morphology.

The domain of ANN approaches (as used here) is that of investigating geometrically meaningful comparisons between forms. Since no biologically grounded decisions are made with respect to which regions of the morphology need to be tracked or otherwise emphasized, biological information is not input into the ANN analytic design in the manner of a morphometric investigation. Rather, biology may be input via the selection of individuals composing each group-specific training set. The word 'may' is used in the previous sentence advisedly. Artificial neural net systems accept such generalized input that, in a very real sense, biological considerations are beside their point. In this way they are more like pure outline-based morphometric analyses (e.g., Fourier methods, standard eigenshape analysis) in which biologically-based landmark mappings play little or no role. The fact that landmark and outline-based analyses can yield similar results, coupled with recent work on landmark-outline hybrid methods (e.g., Bookstein et al. 1999; MacLeod 1999), suggests both approaches are limited by complimentary deficiencies: morphometric methods are rich in biological meaning, but deficient in overall geometric information content while ANN methods are rich in overall information content, but deficient in biological meaning. A synthesis between the two is not only possible, but highly desirable.

Until such a synthesis is achieved, however, it makes sense to match the available strengths of each approach to the diversity of morphological problems at hand. Morphometrics would appear to be best utilized for the investigation of precise distinctions between small groups of morphologically similar species. In such situations, the strengths of a detailed, geometric analysis based on landmark-to-landmark matchings are difficult to ignore. The morphological scope and degree of automation that can be brought to such analyses can be extended by switching the measurement collection strategy to one based on outlines + landmarks rather than using landmarks (or outlines) in isolation (see Bookstein et al. 1999; MacLeod 1999).

Conversely, ANN approaches appear better suited to the characterization of more generalized distinctions between larger groups of morphologically diverse species and their use in α-taxonomic contexts. In these situations the advantages of the greatly expanded diversity of morphologies that can be included, in addition to more complete automation and greater speed in obtaining correct identifications, are equally clear. Moreover, the fora-miniferal analysis results presented above suggest that the inevitable reduction in identification accuracy induced by the relaxation of close topological correspondence need only be minor and that the cost/benefit ratio for time and effort favours the use ANN approaches even in the case of relatively small samples.

5.1.6.2 *Scope for Synthesis?*

As indicated above, we believe there is considerable scope for synthesis between aspects of the geometric morphometric and ANN approaches. In particular, the advantages of the prior processing of interlandmark distance and landmark data using least-square superposition were impressive. In morphometric contexts, size information need not be lost from the system of measurements through this procedure, but can be tracked along with shape as a separate variable (e.g., centroid size; see Bookstein 1986, 1991). The DAISY/ANN implementation could benefit from inclusion of a similar superposition routine that would ensure greater conformance of the basis images prior to subsampling, thereby ensuring greater levels of topological correspondence across the 32×32 pixel maps.

At the moment, this need is handled by a region-of-interest (ROI) routine that provides users with the ability to outline specific features of the specimens and/or segment the image into distinct specimen and background components. This is presently a somewhat time-consuming process that compromises aspects of the ANN approach (e.g., time spent dealing with each image). By strictly limiting the number of landmarks used as the basis for superposition, though, this strategy should be able to be employed successfully by technicians who have low degrees of taxonomic familiarity with the specimens whose images they are processing. There is also considerable scope for maximizing the distinctiveness of each target set image through image warping, though this would introduce an element of sample dependency to the ANN results. Regardless, superposition and image unwarping offer many advantages in interface design as well as in strictly analytical contexts.

On the morphometric side, there is no reason to suppose that PSOM/n-tuple methods could not be applied to fully morphometric data as easily as they are applied to distance data created from pixel maps. Irrespective of its accuracy when used with high-quality superposed landmark data, LDA (along with other multivariate methods) suffers from a pronounced sample sensitivity. This dependency can be ameliorated in principle by obtaining an adequate sample from the population of interest (see MacLeod, 2002b, for an example). In most cases, though, the results of one analysis cannot be adjusted easily to accommodate the inclusion of new objects in previously defined groups, much less the addition of new groups to the discriminant space. PSOM/n-tuple methods were created to address this issue, which is just as problematic for standard morphometric data analysis techniques as it is for ANNs. Accordingly, their application in fully morphometric contexts must be judged as holding considerable promise.

5.1.6.3 *Further Research Directions?*

For morphometric and ANN approaches, one of the most important needs is for better specification of adequate training set attributes. In the technical literature produced on these methods over the years, scarcely any but the most general statements about the composition and nature of reliable training sets have been made. To be sure, a large body of information on statistical sampling theory exists and this should be consulted more often. Nevertheless, training set composition embodies several unique aspects of sample design that have not been explored to date in any systematic manner.

Two aspects of the training set composition issue well illustrated by the foregoing foraminiferal analyses in the context of morphometric approaches are those of specimen orientation and landmark specification. As was noted in the Materials and Methods section, no extraordinary efforts were made to correct inconsistent specimen orientation outside the convention of only including specimens positioned in umbilical view. The reason for this was to mimic what was likely to be the image quality standard that would be available to a technician who needed to make rapid identifications with a minimum of specimen handling. At the outset of our investigation, it was expected that this inconsistency would introduce a measure of error to the results that could compromise some proportion of the identifications.

Similarly, no extraordinary measures were used to ensure that landmark locations were taken at precisely the same locations relative to the underlying morphologies. Rather, these admittedly broad location concepts were 'eyeballed' in quickly with the emphasis on collecting these data as quickly, rather than as carefully, as possible. Despite these consciously inexacting standards, all LDA analyses returned high-quality results — especially those that employed superposition as a preprocessing step. This leads us to suspect that, while no one should ever advocate imprecision as a desirable goal, slavish and time-consuming devotion to absolute minimization of orientation and digitizer error is not required in order to obtain useful results, at least in the context of planktonic foraminiferal species identification.

For ANN approaches, the investigation of training set composition needs is different and, if anything, even broader in scope. Owing to the more generalized types of data that may be used in such systems, an opportunity exists to explore strategies for creating training sets that cover more than a single view of each specimen. For example, a training set could, in principle, be constructed such that it included images of specimens in the standard umbilical, spiral and edge views. Given sufficient distinction between species included in the training set, this may make it possible to construct multiview models of within-species variation and use of these to identify specimens regardless of the orientation a specimen presents to the camera. Additionally, studies seeking to quantify the relation between training set size and identification accuracy for unknown specimens will be important in order to provide more information about the most likely identification for ambiguously determined specimens. Indeed, the entire issue of posterior probability estimation will likely need to be revisited in the context of ANN discriminations, as will the power of different classification algorithms in the identification of different shape classes.

5.1.6.4 *Status within the Systematics Community?*

Throughout this study we have been struck by the negative reception the concept of automated species recognition attracts from many of the taxonomists it is ultimately designed to aid (see also Gaston and O'Neill, in press). Typical objections include allusions to automated systems being too error prone, too complex, too expensive, too slow, and so forth. In many discussions there is also a concern expressed that resources devoted to the development of such systems are wasted and would be better spent training and paying real taxonomists.

Through this investigation, we have attempted to address empirically a number of these concerns. Systems that can authoritatively achieve consistent, semi-automated and fully automated identifications of planktonic microfossil species — and, by extension, many other types of species — to an accuracy of better than 90% over a time frame that ranges from approximately double (LDA) to a small fraction (ANN) of the time it would take a human specialist to accomplish the same task can be constructed at modest expense using available technology. Should this technology become embedded within a distributed, public access computing environment (e.g., the Internet, local intranets), the systematics community would gain a powerful argument for making additional systematic information available throughout academic, public and industrial sectors. Such systems would represent critical, value-added components to already planned international database initiatives and would go a long way to meeting the challenges posed by the taxonomic impediment successfully.

In addition to these considerations, however, a move toward placing automated species recognition at the strategic centre of twenty-first century systematics would have many additional and direct benefits to the science of systematics. The more obvious of these are:

- *Improved access to research funding.* Most research councils (e.g., NSF, NERC, BBSRC, EPSRC, EU) have established interdisciplinary science as the cornerstone of the funding strategies for the foreseeable future. There is also a decided preference for 'big science' as opposed to individual investigator projects. Automated species recognition projects require an interdisciplinary approach and, while they can be pursued at the small-group level, lend themselves to the assembly of large groups of diverse specialists working toward a common aim. At the very least, funding sources for engineering, mathematics and computer science projects could become targets for teams that include a substantial systematics component.
- *Improved ability to take on large-scale biodiversity projects.* A major factor holding back the development of large-scale systematics projects (e.g., biodiversity surveys, synoptic revisions of taxonomy) is the lack of adequate time and manpower to perform to necessary taxonomic identifications to a high degree of accuracy. Automated species recognition projects can play a substantial role in making such projects tractable and fundable.
- *New source of information regarding taxonomic characters.* Systematics has long acknowledged a need for the constant discovery of new characters and character states for use in correctly and consistently recognizing species, populations, etc. At the moment, this process of character/character state discovery is pursued through qualitative approaches yielding decidedly mixed results (e.g., MacLeod 2002b). Automated species recognition systems can operate in authoritative or interactive modes. In this latter context they can become partners with human specialists in systematic research guiding the discovery and testing of new characters and refining the understanding of old characters.
- *Reinvigoration of the discipline of morphological systematics.* In the face of challenges such as DNA bar coding and GeneBank, morphological systematics must become more automated and efficient or it will cease to exist outside a few irreducibly morphology-based refugia (e.g., palaeontology). Because of their generality, automated species (= image) recognition systems can be used in a wide variety

of contexts to integrate different data types and facilitate their combined analysis. This ability extends across the spectrum of systematic data (e.g., morphology, ecology, geography, stratigraphic, chemical, molecular, audio, olfactory, DNA barcodes, SDS protein gel images) and extends well into the quasi-systematic and non-systematic realms.

5.2   SUMMARY AND CONCLUSIONS

In his 1993 review of the status of palaeontological systematics, Roger Kaesler characterized the pros and cons of expert systems and human expertise as shown in Table 2 in his paper. Since 1993, expert systems, in the form of automated species recognition systems, have made significant strides to address several of their deficiencies while losing none of their inherent advantages. Human expertise in taxonomic identification, on the other hand, while not being in any way degraded in principle, has become rarer in the sense that each passing year sees more experienced taxonomists retiring or otherwise becoming unavailable while fewer students — none with the synoptic knowledge gained over a lifetime's engagement with taxonomic issues — step up to take their places. At the same time funding for taxonomic research projects is diminishing, morphological systematics training programmes are closing, and bright students are being attracted into other specialties or leaving the field altogether. One positive way to address this situation is to do what human beings have done ever since the Industrial Revolution when faced with a high-volume and complex, but repetitive, task that needs to be done quickly, consistently and correctly: automate.

A demonstration analysis involving 202 specimens of modern planktonic foraminifera drawn from seven species has shown that traditional distance-based LDA, superposed distance-based LDA, superposed landmark LDA and PSOM/n-tuple ANN approaches can all construct better than 90% correct and consistent discriminant spaces for use in the identification of unknown specimens. Performance of the LDA approach is substantially improved when used in conjunction with superposed landmark data, even when data are collected rapidly from inconsistently oriented, low-quality images in a single orientation. The LDA approach suffers, however, from being semi-automatic, time consuming, labour intensive and working best when all training set objects are morphologically similar.

The PSOM/n-tuple ANN approach can be fully automated, is very time efficient and can be used with a very large diversity of morphologies, but appears marginally less accurate (6.0%) owing to its reliance on gross pixel mapping, which is, in turn, the source of its analytic flexibility. This having been said, LDA and ANN approaches represent substantial improvements in terms of accuracy and consistency over human expertise where experiments show identification reproducibilities can be as low as 30% or lower.

Future developments of LDA and ANN approaches can benefit from cross-fertilization in several areas, especially use of superposition/image unwarping methods to standardize ANN training set images and use of PSOM/n-tuple methods to construct discriminant spaces based on morphometric data. Given the very positive result of our initial investigation of this topic, we see considerable promise in pursuing such development. Overall, it is to be hoped the systematics community will come to appreciate the potential of automated species identification systems to address a number of outstanding problems in systematic theory and practice.

ACKNOWLEDGEMENTS

The idea for this contribution was formulated by the senior author during a public debate on new directions in foraminiferal research held at the FORAMS 2002 Conference in Perth, Australia. NM extends his special thanks to R.K. Olsson and I. Premoli-Silva for the inspirational quality of their comments on automated species recognition methods. Discussion of the relation between geometric morphometric and ANN approaches also benefited greatly from written comments supplied by F. L. Bookstein, who kindly and comprehensively reviewed a previous draft. Funds in support of this study were provided by a Museum Research Fund grant from The Natural History Museum, London.

REFERENCES

Bishop, C.M. (1995) *Neural networks for pattern recognition.* Clarendon Press, Oxford.

Blackith, R.E. and Reyment, R.A. (1971) *Multivariate morphometrics.* Academic Press, London.

Bolli, H.M. and Saunders, J.B. (1985) Oligocene to Holocene low latitude planktic foraminifera. In *Plankton stratigraphy*, ed. H.M. Bolli, J.B. Saunders, and K. Perch-Nielsen. Cambridge University Press, Cambridge, 155–262.

Bollmann, J., Quinn, P.S., Vela, M., Brabec, M., Brechner, S., Cortés, M.Y., Hilbrecht, H., Schmidt, D.N., Schiebel, R., and Thierstein, H.R. (2004) Automated particle analysis: Calcareous microfossils, ETH, Zurich (http://n.ethz.ch/student/bolle/publications/Bollmann2003DERP.pdf).

Bookstein, F.L. (1986) Size and shape spaces for landmark data in two dimensions. *Statistical Science* 1: 181–242.

Bookstein, F.L. (1991) *Morphometric tools for landmark data: Geometry and biology.* Cambridge University Press, Cambridge.

Campbell, N.A. and Atchley, W.R. (1981) The geometry of canonical variate analysis. *Systematic Zoology* 30: 268–280.

Culverhouse, P.F., Williams, R., Reguera, B., Herry, V., and González-Gils, S. (2003) Do experts make mistakes? *Marine Ecology Progress Series* 247: 17–25.

Gaston, K.J. and May, R.M., Taxonomy of taxonomists, *Nature* 356, 281–282, 1982.

Gaston, K.J. and O'Neill, M.A. (in press) Automated species identification — Why not? *Philosophical Transactions of the Royal Society of London.*

Ginsburg, R.N. (1997) An attempt to resolve the controversy over the end-Cretaceous extinction of planktic foraminifera at El Kef, Tunisia, using a blind test. Introduction: Background and procedures. *Marine Micropaleontology* 29: 67–68.

Godfrey, H.C.J. (2002) Challenges for taxonomy. *Nature* 417: 17–19.

Kaesler, R.L. (1993) A window of opportunity: Peering into a new century of paleontology. *Journal of Paleontology* 67: 329–333.

Kennett, J.P. and Srinivasan, S. (1983) *Neogene planktonic foraminifera: A phylogenetic atlas.* Hutchinson Ross, Stroudsbourg, PA.

Lang, R. and Warwick, K. (2002) The plastic self-organizing map. In *World conference on computational intelligence.*

Lucas, S.M. (1997) Face recognition with the continuous *n*-tuple classifier. In *British machine vision conference*, 222–231.

MacLeod, N. (1998) Impacts and marine invertebrate extinctions. In *Meteorites: Flux with time and impact effects*, ed. N.M. Grady, R. Hutchinson, G.J.H. McCall, and D.A. Rotherby. Geological Society of London, London, 217–246.

MacLeod, N. (1999) Generalizing and extending the eigenshape method of shape visualization and analysis. *Paleobiology* 25(1): 107–138.

MacLeod, N. (2000) *PaleoBase: Macrofossils* (part 1). Blackwell Science and The Natural History Museum, Oxford.

MacLeod, N. (2002a) Morphometric perspectives on the MorphoBank project. In *Sixth international congress of systematic and evolutionary biology*, ed. M. Michevitch. University of Patras, Patras, Greece, 196.

MacLeod, N. (2002b) Phylogenetic signals in morphometric data. In *Morphology, shape and phylogeny*, ed. N. MacLeod and P.L. Forey. Taylor & Francis, London, 100–138.

MacLeod, N. (2003) *PaleoBase: Macrofossils* (part 2). Blackwell Science and The Natural History Museum, Oxford.

MacLeod, N. and Forey, P. (2002) *Morphometrics, shape and phylogenetics*. Taylor & Francis, London.

Marcus, L.F., Bello, E., and García-Valdecasas, A. (1993) *Contributions to morphometrics*. Museo Nacional de Ciencias Naturales 8, Madrid.

Marcus, L.F., Corti, M., Loy, A., Naylor, G.J.P., and Slice, D.E. (1996) *Advances in morphometrics*. Plenum Press, New York.

Neff, N.A. and Marcus, L.F. (1980) *A survey of multivariate methods for systematists*. Privately published, New York.

Pimentel, R.A. (1979) *Morphometrics: The multivariate analysis of biological data*. Kendall/Hunt, Dubuque, IA.

Reyment, R.A., Blackith, R.E., and Campbell, N.A. (1984) *Multivariate morphometrics*, 2nd ed. Academic Press, London.

Ripley, B.D. (1996) *Pattern recognition and neural networks*. Cambridge University Press, Cambridge.

Rohlf, F.J. and Bookstein, F.L. (1990) *Proceedings of the Michigan morphometrics workshop*. The University of Michigan Museum of Zoology Special Publication 2, Ann Arbor.

Schalkoff, R.J. (1997) *Artificial neural networks*. MIT Press and the McGraw–Hill Companies, Inc., New York.

Sokal, R.R. and Sneath, P.A. (1963) *Principles of numerical taxonomy*. W.H. Freeman, San Francisco.

Weeks, P.J.D., Gauld, I.D., Gaston, K.J., and O'Neill, M.A. (1997) Automating the identification of insects: A new solution to an old problem. *Bulletin of Entomological Research* 87: 203–211.

Weeks, P.J.D., O'Neill, M.A., Gaston, K.J., and Gauld, I.D. (1999a) Automating insect identification: Exploring limitations of a prototype system. *Journal of Applied Entomology* 123: 1–8.

Weeks, P.J.D., O'Neill, M.A., Gaston, K.J., and Gauld, I.D. (1999b) Species identification of wasps using principle component associative memories. *Image and Vision Computing* 17: 861–866.

Wheeler, Q.D. (2003) Transforming taxonomy. *The Systematist* 22: 3–5.

Wheeler, Q.D., Raven, P.H., and Wilson, E.O. (2004) Taxonomy, impediment or expedient? *Science* 303: 285.

Young, J.R., Kucera, M., and Cheng, H.-W. (1996) Automated biometrics on captured light microscope images of coccoliths of *Emiliana huxleyi*. In *Microfossils and oceanic environments*, ed. A. Moguilevsky and R. Whatley. University of Wales, Aberystwyth, Wales, 261–277.

Zachariasse, W.J., Riedel, W.R., Sanfilippo, A., Schmidt, R.R., Brolsma, M.J., Schrader, H.J., Gersonde, R., Drooger, M.M., and Brokeman, J.A. (1978) Micropaleontological counting methods and techniques — An exercise on an eight-meter section of the Lower Pliocene of Capo Rossello, Sicily. *Utrecht Micropaleontological Bulletins* 17: 1–265.

6 Automated Extraction of Biodiversity Data from Taxonomic Descriptions

Gordon B. Curry and Richard J. Connor

CONTENTS

ABSTRACT

Taxonomic descriptions are the core output of systematics research and of critical importance for key questions in the fields of biology, earth science and environmental science. These descriptions contain vast amounts of information about the morphological features of organisms on Earth, their geographic distribution and, for fossils, their geological history. Much of these data are not widely available to the many potential users because they are predominantly published as hard copy in systematics journals or monographs. Digitization of these descriptions would make them much more widely available, but doing this manually would be an enormous and unrealistic task. This chapter describes an alternative method of automating the digitization of taxonomic descriptions, using new techniques in computing science that exploit the high degree of structure and organization imposed by systematic convention and rigorous editorial procedures. The method involves parsing such partially structured text to generate XML tags around discrete sections of the text. Once tagged, complex queries can be run across the data that were not possible with the non-tagged text, and the tagged text can more readily be imported into an existing database if required. A major bottleneck in the construction of biodiversity databases would therefore be overcome if the extensive data present in taxonomic descriptions could be extracted by computer and not rely on human operators manually entering the information into database fields. The advantages of automating the data capture phase of biodiversity database development are numerous — the process is fast, flexible in terms of input data and output data, accurate and can readily be updated. Adopting this strategy would mean that computers are doing the boring repetitive part of the process for which they are ideally suited, freeing humans to devote more time and resources towards the creative, analytical exploitation of these data. Issues such as copyright and intellectual property need to be addressed, but these are well within the capabilities of the kinds of cyber-infrastructure being developed in computing science. It also suggests that museums and other repositories of natural history collections should urgently review their policies on the publications of taxonomic descriptions based on specimens in their collections. In the digital world, it may well be that digitized data from collections-based research should be managed and maintained every bit as assiduously as the specimens are. An obvious way forward would be to adopt a twofold strategy of preparing XML templates for future taxonomic descriptions that allows synchronous publication and digital captures, and a separate phase of scanning, digitization of existing taxonomic monographs (many of which are full of relevant and beautifully illustrated taxonomic data).

6.1 INTRODUCTION

Systematics is widely acknowledged as an essential core discipline that underpins all branches of biology and palaeontology. Yet recent years have seen a major decline in the subject, despite the fact that it has a central role in current major scientific issues such as biodiversity, climate change, evolution and human health. This problem in systematics is a global phenomenon and has been the subject of many investigations and reports. The message is unequivocal: systematists are retiring and not being replaced, biodiversity collections are being neglected and major groups of organisms are not being investigated due to the lack of suitable expertise. Various initiatives have attempted to address this problem, but it remains clear that systematics is in crisis.

At least one aspect of the problem facing systematics is that much of its output remains unavailable to the majority of potential users. The core products of systematics research are taxonomic descriptions, which are generally published in specialist journals or monographs, using language that is difficult for the non-systematist to utilize. Moreover, the vast majority of data are only available as printed documents and are not yet available in a digital format; this is a further impediment to wider distribution and utilization. In conjunction with the fact that most systematists work individually or in small groups, it is not surprising that the discipline is seen as small, piecemeal science at a time when attention is focused on 'big science' subjects such as molecular biology and e-science.

This chapter argues that systematics is undoubtedly big science when taxonomic descriptions are considered collectively, rather than in isolation. The challenge is to make this information much more widely available to users in the wider scientific community, government and the general population. As discussed later, new techniques in computing science do offer a method of doing this, without compromising the scientific rigor of systematics research.

The specific problem addressed in this chapter is of generating digital data from taxonomic descriptions. The normal method of doing this is by entering the information into a database, which is a daunting and often unrewarding task when carried out manually. Indeed, digitization of biodiversity data, including taxonomic descriptions, is widely seen as a major bottleneck in the development of a digital biodiversity network. However, generating such digital data automatically using suitable computer software is now a real possibility, due to attributes of taxonomic writing that readily lend themselves to automation.

6.2 SCOPE OF THE DATA

The range of information available from taxonomic descriptions is much wider than just the binomial name of a taxon. The great value of these data stems from the fact that they contain information about morphology of the organism, including those features that are diagnostic for this taxon. Standard descriptions also give information on authorship, synonyms (previous names applied to this taxon) and location and provenance of the type or illustrated specimens. Biogeographical information is provided in the form of localities and ranges and, for palaeontological descriptions, there is also information on the stratigraphical range of the taxon.

Many illustrations are of extremely high quality in taxonomic manuscripts and are extremely valuable for taxonomic research, provided that they can be readily accessed. The advantages of digitizing such information is that much of the data are not likely to change; taxonomic names may be revised, but the morphological features are likely to remain valid. Digitization of taxonomic descriptions should be seen as a critical component of the global biodiversity network of digitized information; the information they contain includes much valuable information not included in biodiversity databases (often containing only species names) or digitized museum records (often listing only taxonomic name, location, type, etc. without morphological features).

The wide range of information contained in each taxonomic description makes their digitization of widespread interest. Many scientific questions that are currently difficult or impossible to answer could be addressed if all the information present in taxonomic descriptions were available in databases. The biogeographical distribution of large numbers of taxa or groups of taxa could be investigated much more readily and flexibly from databases than if the data had to be extracted manually. In addition, it would be possible to investigate the evolution and distribution of morphological features and to compare these in different groups. Collating the lists of authors and synonyms would provide an invaluable historical perspective on the development of phylogenetic interpretations and provide a reliable monitor of the state of systematics effort. With an average of 20–30 or more separate bits of information available for each taxon in a database, there are numerous ways for the data to be explored to reveal new insights into evolution and biogeography. The resulting databases would be major resources for biodiversity, conservation and climate change research.

6.3 HISTORICAL LEGACY OF SYSTEMATICS

For most groups of organisms, there is an extensive taxonomic literature extending back for hundreds of years. While some of the really early work in this field may have been superceded by subsequent research, there is an extensive and enormously valuable literature extending back for several centuries. Monographs written over 100 years ago utilize layouts and terminologies that are still in use today, and they are often important sources of information for present-day systematists. The taxonomic name assigned to a particular species may have changed since the publication of these ancient descriptions, but the morphological descriptions remain valid and could easily be cross-referenced using the synonym list for each species. In reality, however, many of these ancient taxonomic monographs are rare documents and may only be available in specialist libraries.

There is another major reason for digitizing the rich legacy of taxonomic monographs and making the information they contain much more widely available. These monographs provide a unique overview of the historical pattern of changing biodiversity, which is a major topic of interest at the present time. Along with well-documented specimens held in museum collections, historic taxonomic monographs provide a reliable method of assessing how environmental factors, such as climate change or global warming, have affected the distribution of organisms on Earth. It is abundantly clear that there have been huge changes in biodiversity in recent decades (e.g., Hawkesworth 2001) attributed to environmental change, human impact, imported species and a variety of other global and regional events. Documenting and understanding such changes represents a major scientific challenge, and documenting historic taxonomic monographs would provide a massive reservoir of relevant information, much of which is currently not exploited because of its inaccessibility.

It is clear that the often standardized layout of formal taxonomic descriptions was developed many years ago. In a number of areas that are important for digital data processing, the core conventions of taxonomic descriptions have remained unchanged up to the present day. A good example of this is a comparison between the description of the brachiopod *Atretia gnomon* (Jeffreys) (now known as *Cryptopora gnomon*) published in a seminal work on Brachiopoda by Davidson (1887) and the generic diagnosis of *Cryptopora* as given in the brachiopod treatise (Williams et al. 2002). The equivalent terms used in the two publications are shown in Table 6.1, and without exception they are directly equivalent to one another. Although there have been numerous changes in the classification of brachiopods in the intervening 110 years (Williams et al. 1965, 2002), the morphological descriptors applied are virtually identical or at least sufficiently similar to be readily recognizable to systematists.

The fact that the names have changed is relatively insignificant in the context of taxonomic databases because the two taxa are easily cross-referenced by virtue of their synonym lists and type species citation. These ancient monographs clearly represent a huge store of high-quality taxonomic data and accompanying illustrations, and they represent a major resource for biodiversity if the data and illustrations could be made available digitally. The antiquity of many such monographs means that copyright issues may not apply, but the hurdle of manually entering the taxonomic data remains.

As discussed in following text, modern digital scanning techniques and optical character recognition (OCR) software have advanced considerably in recent years, and producing digital versions of the text and illustrations from ancient taxonomic monographs is now achievable in a realistic timescale. To investigate this possibility, parts of the Davidson monograph were scanned and the text input into an HTML (Web format) page. The results clearly show that digitization of taxonomic monographs is viable with minimal work. Best results were achieved when the OCR had a dictionary of taxonomic terms used in the text. This indicates that scanning is most accurate when conducted by a taxonomist familiar with the range of specialist and technical terms used in the taxonomic description. It is

TABLE 6.1

Comparison of Terms Used in Description of *Atretia gnomon* Jeffreys by Davidson, 1887, and of the Same Species (now Known as *Crytopora gnomon* [Jeffreys]) in the *Treatise on Invertebrate Paleontology*[a]

Davidson, 1887, p. 173	Treatise, 2002, p. 1244
…Very small, triangularly oval or pear shaped…	Small…subtrigonal to ovoid lenticular
Dorsal valve slightly convex…ventral valve slightly deeper	…Almost equibiconvex
…With wide slightly raised medial fold	…Rectimarginate to broadly sulcate
Surface smooth	…Smooth
…Beak moderately produced…	…Beak moderately long, pointed
Angular at its extremities	Nearly straight
…Narrow rudimentary deltidial plates…	…Deltidial plates auriculate
…Triangular incomplete foramen…	Rudimentary, disjunct
…Dorsal…mesial septum…	…Dorsal median, septum high
Large vertical blade-like plate	
…Small cardinal process	…Cardinal process small and transverse
…Two short, slender, curved lamellae denticulated at their extremities	…Crurae digitate distally
…Short but strong diverging dental plates	Dental plates disjunct, subvertical

[a] *Part H, Brachiopoda*, vol. 4, Geological Society of America and University of Kansas, 2002.

likely, however, that once a comprehensive dictionary of terms is available, the scanning process could be carried out by less experienced personnel. Best results were achieved by first preparing photocopies of the monograph pages. The intensity of photocopy reproduction was reduced to minimize the interference caused by blemishes, stains, etc. in printed pages that were over 100 years old (Figure 6.1).

6.3.1 How to Get Information from Taxonomic Descriptions into a Database

While there are many good scientific reasons justifying the digitization of taxonomic monographs, there are many hurdles in actually implementing such a programme in present-day terms. Undoubtedly database research has developed rapidly in recent years, as have been the analytical protocols required to investigate the data they contain. The main problem facing efforts to digitize taxonomic descriptions is how to get the information into the database in the first place. Currently, this is done manually, with the data being typed into a computerized database (Figure 6.2). This is a huge task, even for small numbers of taxa, and suffers from a number of major problems:

- Data entry is an unrewarding, repetitive task.
- Information often has to be extracted from the text and recoded to fit within the fields of the database.
- If more than one person is involved in data entry, there can be variations in how data are extracted and recoded between individual operators.
- Retyping information inevitably results in mistakes.
- It is virtually a never-ending task because it will be necessary to re-enter data as taxonomic revisions occur.

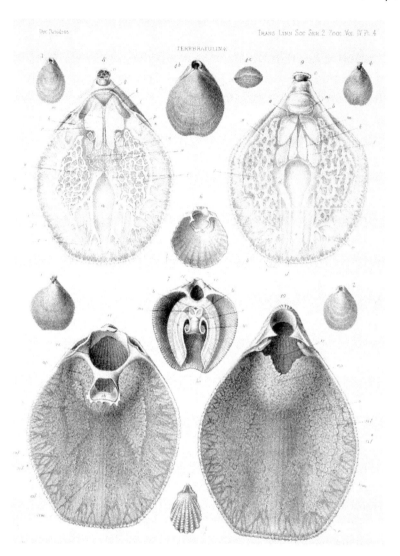

FIGURE 6.1 Sample of the high quality illustrations provided in historical taxonomic monographs. (Davidson, 1887)

The fact that so few taxonomic descriptions are available in digital format is testimony that it is difficult to find the necessary human and financial resources to allow data entry on such a huge scale. There are numerous biodiversity databases, but they mostly do not contain the full range of information present in taxonomic descriptions. What is needed is to reverse the process and use the computer to do the boring, repetitive work of data entry and hence allow humans to concentrate on the creative, interpretative aspects. In theory, computers are much more suited to automatic digitization of data than humans, but in practice, suitable software has not been available to allow this to happen. Recent developments in computing science (in particular, the processing of semistructured text) have opened up the prospect of being able to automate the extraction of taxonomic data. This chapter describes how the inherent characteristics of taxonomic writing can be exploited to overcome this major impediment to the digital exploitation of taxonomic information.

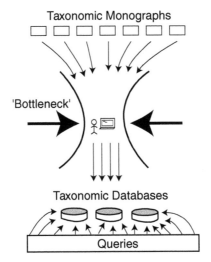

FIGURE 6.2   Schematic diagram illustrating the bottleneck resulting from the need to manually enter taxonomic information into the database.

6.4   BASIC PROBLEM

The main reason for creating a database is to organize the available information into a number of discrete, labelled fields, which can then be searched to provide the information required by the user (Figure 6.3). Human-readable text is not structured according to logical fields and is seen by the computer as a string of characters, preventing the automation of complex queries. A database, on the other hand, is essentially a collection of separate fields, allowing the accurate specification of complex queries. An example (based on a palaeontological description) of the sort of query facilitated by a database would be to extract the subset of all taxa that had strongly biconvex shells with spines on the exterior surface that occurred during the Ordovician period and had been collected or recorded from the USA.

Taxonomic descriptions contain all this information and much more. However, computer-based understanding of natural language does not yet and may never allow the automation of such queries over text resources. Querying text using these criteria could locate places in the text where the words 'strongly biconvex', 'spines', 'Ordovician' and 'United States' were cited in close proximity, but understanding the meaningful connections among the terms cannot be achieved; even finding synonyms and misspellings of these terms is problematic. Thus, extensive human involvement and expertise would be required for each such query.

Simple searches of text strings are in any event inadequate for the purpose because a certain amount of important information is implied rather that directly stated in standard text. The stated stratigraphic range of Cambrian to Devonian, for example, indicates that the taxon had a stratigraphical range that did indeed include the Ordovician (one of the intervening geological periods between the Cambrian and the Devonian). The fact that it is not directly mentioned in the text means that simple queries will miss this taxon when searching for Ordovician. Databases can handle this by having separate fields for the start and end of a stratigraphic range (e.g., filled by Cambrian and Devonian, respectively) and by using a look-up table to allow a complex search that will successfully extract taxa whose stratigraphic range includes the Ordovician, even though the term does not appear in the

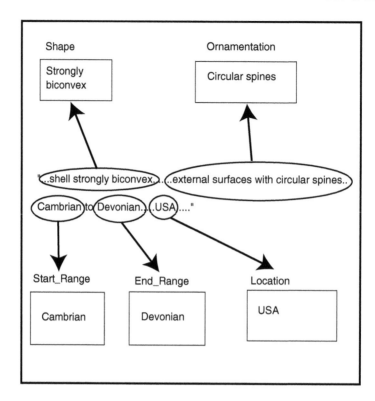

FIGURE 6.3 Diagram showing how components of a taxonomic descriptions have to be partitioned into labeled boxes (fields) when imported into the database.

original description. This still leaves the problem of how to get the information into the fields of the database.

6.4.1 THE SPECTRUM FROM NONSTRUCTURED TO STRUCTURED ELECTRONIC DATA

In computing terms, the fundamental difference between normal text and databases is described in terms of structure. Normal text is unstructured, while the labelled fields of a database represent terms that have a high degree of structure. Using the preceding example, by transforming the unstructured text of taxonomic descriptions into the structured terms of a database, it would be possible to successfully carry out a complex search of the latter that would return, in a single document, all the taxa from a particular superfamily that had spines and were found in the Ordovician of China. The question is whether or not it is possible to achieve the transformation from unstructured to highly structured text without having to depend on having the text retyped by a human operator.

Having highly accessible electronic data resources is the key for the future health of any big science and has been explicitly identified as the crucial future direction for systematics. However, the overall design of such resources requires careful thought because large digital data collections may be structured in a range of different ways. It is important to clearly identify which method is most suitable for a particular purpose and, indeed, which methods are feasible for particular types of information.

Recent developments in computing science have provided a much wider range of methods of obtaining structured digital data. In this context, it is useful to think in terms of a spectrum of data structuring, which ranges from totally unstructured data at one end, up to a highly structured database at the other. The two end points are well recognized amongst the users and creators of taxonomic literature and biodiversity databases; a single large text resource, such as a taxonomic monograph, is an example of an unstructured digital resource, and the database is a good example of a highly structured digital resource. However, there are various possible hybrids along this line also, one of which in particular will be advocated in this chapter as the most suitable for important collections of taxonomic data. There is great value for systematics in exploiting this spectrum of approaches, especially because it holds out the possibility of greatly speeding up the acquisition of a digital data resource for the subject.

The unstructured text resource is the simplest to acquire because text is the traditional format for all current and historical publication. While modern publications typically exist in electronic format already, if this is made available by its owners, it is also a relatively straightforward task to create electronic text resources for archival and indeed ancient material, due to recent significant advances in document scanning and OCR technology (Figure 6.4). Together, these techniques make it eminently feasible for large volumes of archive text to be brought into the electronic domain at relatively small human cost.

Nor should unstructured text be dismissed as an information format; advances in automated *information retrieval* and *information extraction*, spurred by their requirement to improve the efficacy of Internet search engines, mean that pertinent information can easily and quickly be found in a very large document resource, even modulo details such as differences in the textual form of terms or even the use of synonyms. As any user of an Internet search engine will know, these techniques give excellent performance in *precision* and *recall* (terms meaning, roughly, specificity and sensitivity, respectively), even over a document collection in the order of 10^9 resources. However, in terms of extracting rigorously correct answers to precisely defined queries, the probabilistic essence of the approach leaves much to be desired; furthermore, the result of an information retrieval engine will be a set of resources requiring inspection by a human, rather than those that could be used as the input to further automated query processing. As mentioned previously, queries using text strings alone will not satisfy the majority of questions that users of systematics resources require from digital resources, but they may offer a useful facility for certain purposes, such as initial investigations of the data.

At the other end of the spectrum is the database, which contains a highly regular subset of all the available information. Databases have the significant advantage of providing a framework where rigorously defined automated queries can be posed, and very high confidence may be vested in the answers. The reason for this high degree of confidence lies in the database schema, which is an *a priori* data description designed for the particular set of tasks in hand. This schema is used to ensure that queries are *sound* — that is, that they strongly correspond with the description of the data used. For example, a question of mean age may only be posed if every record has an age field and each one of these is a correct numeric value. The presence of the schema implies that the property of soundness may be determined irrespective of the actual data collection, by reference to the schema alone. If a query is not sound, the programmer will be immediately alerted and the query will not be allowed to execute.

Furthermore, the database is also populated with information according to the structure of this schema, giving an enforced quality filter at data entry time. If, for example, the field

2. LIOTHYRIS ARCTICA, Friele, sp. (Plate I. figs. 17, 18.)

Terebratula arctica, Friele, Særskilt Aftryk af Nyt Magazin for Naturvidenskaberne, pl. i. fig. i., 1877.

Shell small, globose, broadly ovate, rather longer than wide. Valves smooth, glassy, semitransparent, whitish; dorsal valve convex, squarely circular, without fold or sinus; ventral valve very convex and deep; beak unusually short, slightly incurved and truncated by a very small foramen margined anteriorly by rudimentary deltidial plates; loop very small and simple. Length 7, breadth 6, depth 4 lines.

Hab. Dredged by Herman Friele some few miles south-west of Jan Mayen, in 263 fathoms depth. Shell abundant, but so brittle that most of the specimens were broken during the dredging-operation.

Obs. After having carefully compared a specimen of the shell under description, sent to me by Friele, with others of the var. *minor* to which it had been referred by Dr. Jeffreys, I could, as Friele had previously done, discover several differences which, although not very great, have induced me to follow its discoverer in considering it a distinct species. *L. arctica* is much more globose and squarely rounded than *L. minor*, which is more of an elongated oval. As stated by Friele, its form approaches most to *L. minor* of Philippi, but the deviation is shown in the shorter beak and by the position of the foramen, which, in *L. arctica*, is placed directly above the dorsal valve, the deltidium being almost hidden. The loop in *L. arctica* is very much weaker and thinner, and the crura processes are placed further apart than in *L. minor*. It is the first representative of the genus *Liothyris* that has been hitherto found in Arctic seas.

2. LIOTRYRIS ARCTICA, Friele, sp. (Plate I. figs. 17, 18.)

Terebratula arctica, Friele, Srerskilt Aftryk af Nyt Magazin for Naturvidenskaberne, pl. i. fig. i., 1877.

Shell small, globose, broadly ovate, rather longer than wide. Valves smooth, glassy, semitransparent, whitish; dorsal valve convex, squarely circular, without fold or sinus; ventral valve very convex and deep; beak unusually short, slightly incurved and truncated by a very small foramen margined anteriorly by rudimentary deltidial plates; loop very small and simple. Length 7, breadth 6, depth 4 lines.

Hab. Dredged by Herman Friele some few miles south-west of Jan Mayen, in fathoms depth. Shell abundant, but so brittle that most of the specimens were during the dredging-operation.

Obs. After having carefully compared a specimen of the shell under description, sent to me by Friele, with others of the var. *minor* to which it had been referred by Dr. Jeffreys, I could, as Friele had previously done, discover several differences which, although not very great, which have induced me to follow its discoverer in considering it a distinct species. *L. arctica* is much more globose and squarely rounded than *L. minor*, which is more of an elongated oval. As stated by Friele, its form approaches most to *L. minor* of Philippi, but the deviation is shown in the shorter beak and by the position of the foramen, which, in *L. arctica* is placed directly above the dorsal valve, the deltidium being almost hidden. The loop in *L. arctica* is very much weaker and thinner, and the crura processes are placed further apart than in *L. minor*. It is the first representative of the genus *Liothyris* that has been hitherto found in Arctic seas.

FIGURE 6.4 OCR-scanned text from taxonomic description in Davidson, 1886. (a) = original text; (b) = OCR processed text from original manuscript. (Davidson, T.D. *Transactions of the Linnean Society of London*, 4, Linnean Society, 1887)

entered does not correspond to the expected value range or if a field fails to be entered through oversight, a notification will immediately occur and the error must be rectified at that point. Finally, database constraints may be defined so that any inconsistent changes to the database will cause an alert. All of these valuable properties are possible due to the presence of the database schema; this definition is the first requirement for any new database.

These are the good points of the database; the single worst aspect, however, is that the strong reliance upon the definition of a precisely defined schema necessarily allows only a subset of all the available information to be stored. This is acceptable for many purposes, notably when a fixed set of queries is required over information that is inherently regularly structured; however, in a field where data contain many inherent irregularities, the partial nature of the schema always causes problems with the loss of information. This aspect is aggravated by the fact that the schema, once in place, is extremely difficult to change, precluding smooth evolution as the demands of a system's users evolve over time.

A hybrid data model, known now as the semistructured model, appeared in the database research literature in the late 1990s. Originally proposed as an interchange format (i.e., a partial solution for sharing information between different databases), the realization of the value of its use as a data model in its own right quickly made it a mainstream research topic. The inherent value of semistructured data is that they are self-describing; that is, they contain structure, but this structure is an integral part of the data collection, rather than appearing as a separate structural definition, as in a database schema. Semistructured data are technically defined as an edge-labeled directed graph and are often portrayed using data diagrams such as that shown in Figure 6.5.

The significance of the correspondence between these diagrams and those in common use by taxonomists will not be missed; one of the key advantages of this data model in the present context is that the user population is well used to thinking in terms of this as a natural data model, rather than in terms of relations that have a rather artificial mapping.

This research, details of which are well beyond the scope of this chapter, occurred by historical coincidence around the same time that XML was emerging as a replacement standard for HTML, with the primary purpose of separating the concepts of information and presentation within Internet resources. The XML standard, while containing much historical 'junk' from this origin, also happens to provide a public, open standard encoding for semistructured data.

Semistructured data are sometimes referred to as a potential replacement for databases; this is a mistake. They are purely a compromise model, giving a useful hybrid between free text and databases. While semistructured resources can have data descriptions (the current standards being DTD and XMLSchema), these are very different concepts from database schemata: they are not necessarily defined *a priori*, queries are not necessarily checked for soundness with respect to them and there is no framework for enforcing data to correspond to the descriptions. While these are all ongoing research issues, the essential case is that semistructured data provide more flexibility but less safety than traditional databases.

We should stress again that the semistructured data model should be viewed as a hybrid, rather than as a new model to replace existing ones. In terms of the example queries given before, using a semistructured paradigm is quite good for all of them. It is always better than either structured or unstructured models for things each is bad for; it is never as good as either for things for which each is good.

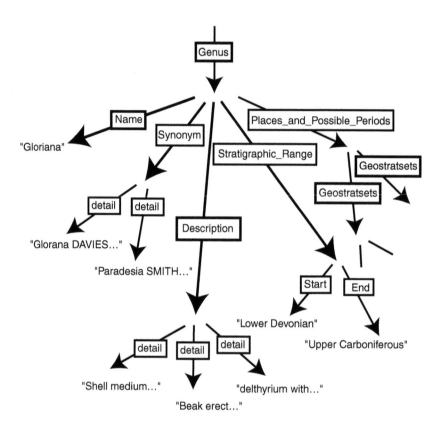

FIGURE 6.5   Tree picture of data.

In this context, the single greatest advantage we perceive is the ability to avoid any data loss. Furthermore, the flexibility of the format is enshrined in the ability to move in either direction from it; for those sets of tasks that do require the rigor of databases, the creation of those databases is made substantially easier by having the semistructured, rather than unstructured, information as its starting point. In the other direction, when free text is the requirement (as, for example, when people just need to read species descriptions as from a traditional treatise), this is just one of the possible views that may be provided by a semi-structured collection.

Our novel observation in this context is that semistructured data may be gleaned, via a largely automated approach, from printed textual archives. This is due to the following:

- Ancient archival material may be successfully subjected to scanning and subsequent OCR to create an electronic text archive.
- This text archive is highly structured due to the use of rigid conventions that have developed within each taxonomic subdomain, and the inherent structure within the document can be gleaned via a further automated analysis of the scanned text.
- The quality of structure that can be gained through these almost totally automated processes is sufficient to perform a large class of automated query over the meaning, rather than just the textual form, of the information.

6.5   CONVENTIONS OF TAXONOMIC DESCRIPTIONS

In the context of this chapter, the key feature of taxonomic descriptions is that they are often written in a standardized format, which determines the order in which the data are presented. For example, most descriptions start with the taxon name, details of the author, the date of publication, where it was published and a page reference, and information about synonyms (names that have previously been applied to that taxon). This would be followed by the description of the characters of the taxon, often again following a standard pattern. In the case of genera described in the brachiopod treatise for example, the description starts with the features of the overall shape and profile of the entire shell and any surface ornamentation. The internal features are then listed, followed by information on the stratigraphic and geographical distribution, and any type or figured material. A similar pattern is discernible in the formal descriptions of most taxa, from plants to bacteria, although the details will vary depending on the characteristics of the taxa and on historical precedent.

The following example demonstrates a typical, but hypothetical, taxonomic description of a taxon, based on the protocols adopted in the *Treatise on Invertebrate Paleontology, Part H* (4) Brachiopoda (Williams et al. 1997, 2002).

Gloriana CHRESSMAN, 1986c, p. 313 [**Cryptiana superbia* NELSON, 1856, p. 413; OD] [=*Paradesia* SMITH, 1871, p. 54 (type *P. excella* JONES 1869b, pl.); = *Glorana* DAVIES, 1906, p. 18]

Shell medium to small, subcircular to elongately oval outline, strongly biconvex. Beak erect to suberect; delthyrium with conjunct delthyrial plates. Anterior commisure rectimarginate; exterior with faint radial ribs, and prominent, concentric growth lamellae, each bearing small, regularly distributed, erect spines. Dental plates absent; cardinal process long, thin, bifurcating anteriorly. *?Lower Devonian (?Emsian), Upper Devonian (Lochkovian), Upper Caboniferous (Cantabrian)*: Australia *?Emsian*; ?England, Belgium, Germany, Czechoslovakia, *Eifelian–Givetian*; England, Ireland, Germany, Russia; *Frasnian–Cantabrian.*

This description, although just over 100 words in length, contains a prodigious amount of information. The headers give information on the suprageneric classification, and the synonyms are listed within the squared brackets. The first sentence in the main paragraph describes the size and shape of this organism, while the second gives detail of the posterior features. The third sentence describes the anterior features and the ornamentation present on the exterior of the shell. The remaining descriptions provide information on features visible inside the organism. The italicized section at the end of the description outlines the known distribution of the taxon in space and time. In this case, it indicates that the taxon's earliest appearance was possibly in the Emsian stage of the Lower Devonian period in Australia. The '?Emsian' citation indicates that there is some doubt about this Australian record, but it is important to retain this information, even if it requires further investigation.

The next section of the stratigraphic part of the description indicates that reliable records indicate that the taxon was definitely present in the Upper Devonian period (from the Lochkovian stage) and survived into the Cantabrian stage of the Upper Carboniferous period (a geological range of approximately 110 million years). The final section of the description documents the geographical locations were the taxon has been discovered, and it is broken up into combinations of stratigraphic and geographic information that provide an overview of the palaeobiogeographic distribution of the taxon.

6.6 AUTOMATIC EXTRACTION OF INFORMATION FROM TAXONOMIC DESCRIPTIONS

The partially structured or semistructured nature of taxonomic descriptions has allowed the development of an automated method of digitizing the data they contain. The procedure involves the development of a parser program that can process text in such a way that it generates XML tags around different segments of the text based upon the standardized stylistic protocols adopted in the taxonomic description. These tags act in a similar way to the fields in databases, by subdividing the text in such a way as to allow complex queries to be run over the text. Unlike the database approach, however, the tags are generated by the software and do not require human intervention.

Figure 6.6 shows the results of running our XML parser across the sample taxonomic description shown above. In the first section of the resulting XML-tagged document, the taxon name has been identified, and the author has been separately labelled, as have the date and page citation of the original description. This is achieved purely on the basis of the style, organization and punctuation of the text — the taxon name in bold, followed by the author's name in capital letters and then the date and the page citation separated by commas.

The parser then correctly identifies and labels the synonyms, which follow the author details, with each synonym enclosed within square brackets. Synonyms are again important information, tracking the history of binomial names applied to this particular taxon. The next section of the taxonomic description provides information on the morphological features of the taxon, represented in this case by a paragraph of approximately 50 words. The key components of the morphological description are separated by full stops or semicolons, allowing the parser to isolate and label these separately as details of the description.

The third major section of the XML document in Figure 6.6 presents the overall stratigraphic distribution of the taxon. The organization of the text indicates that a range is given, so within the stratigraphic information presented, the parser recognizes this as STRATRANGE, and tags both the doubtful starting period of the Lower Devonian and the more reliable Upper Devonian. The confirmed=false attribute distinguishes unreliable from reliable stratigraphic information (shown as confirmed=true). The DETAILPERIOD tag cites the stages within the periods, if these are provided.

The fourth section of the XML document tags the palaeobiogeographical information as a series of 'places and possible periods'. Each GEOSTRAT set first lists the places that the taxon has been found (PLACE) and then the stratigraphic range of the taxon in those places (STRATIGRAPHICPERIOD). Again, the convention of retaining uncertain records but labelling them as confirmed='false' is retained.

6.7 QUERYING XML-TAGGED TEXT

The once concise taxonomic description has been transformed into a much longer, much more unwieldy document by the XML parser, but the important aspect for the automation of digital capture is that that the tags allow complex searches to be carried out of the type

```
- <GENUS confirmed="true">
<NAME>Gloriana</NAME>
 <AUTHOR>CHRESSMAN</AUTHOR>
 <DATE>1986c</DATE>
 <PAGE>p. 313</PAGE>
 <TYPE>[*Cryptiana superbia Nelson, 1856, p.413]</TYPE>
 <SYNONYM>[=Paradesia SMITH, 1871, p. 54 (type, P. excella
JONES, 1869b, pl. 6)]</SYNONYM>
 <SYNONYM>[= Glorana DAVIES, 1906, p.18]</SYNONYM>

- <DESCRIPTION>
 <DETAILS>Shell medium to small, subcircular to elongage oval
outline, strongly biconvex </DETAILS>
 <DETAILS>Beak erect to sub-erect</DETAILS>
 <DETAILS>delthyrium with conjunct delthyrial plates</DETAILS>
 <DETAILS>Anterior commisure rectimarginate</DETAILS>
 <DETAILS>Exterior with faint radial ribs, and concentric growth
lamellae, each bearing small, regularly distributed, erect
spines</DETAILS>
 <DETAILS>Dental plates absent</DETAILS>
 <DETAILS>cardinal process long, thin, bifurcating
anteriorly</DETAILS>
 </DESCRIPTION>

- <STRATIGRAPHIC>
  <STRATIGRAPHICRANGE>
   <START confirmed="false">Lower Devonian</START>
    <DETAILPERIOD confirmed="false">Emsian</DETAILPERIOD>
   <START confirmed="true">Upper Devonian</START>
    <DETAILPERIOD confirmed="true">Lochkovian</DETAILPERIOD>
   <END confirmed="true">Upper Carboniferous</END>
    <DETAILPERIOD confirmed="true">Cantabrian</DETAILPERIOD>
  </STRATIGRAPHICRANGE>
  </STRATIGRAPHIC>

- <PLACES_AND_POSSIBLE_PERIODS>
 - <GEOSTRATSETS>
  <PLACE confirmed="true">Australia</PLACE>
  <STRATIGRAPHICPERIOD
confirmed="false">Emsian</STRATIGRAPHICPERIOD>
  </GEOSTRATSETS>
- <GEOSTRATSETS>
  <PLACE confirmed="false">England</PLACE>
  <PLACE confirmed="true">Belgium</PLACE>
  <PLACE confirmed="true">Germany</PLACE>
  <PLACE confirmed="true">Czechoslovakia</PLACE>
  <STRATIGRAPHICPERIOD>
      <START confirmed="true">Eifelian</START>
      <END confirmed="true">Givetian</END>
  </STRATIGRAPHICPERIOD>
  </GEOSTRATSETS>
- <GEOSTRATSETS>
  <PLACE confirmed="true">England</PLACE>
  <PLACE confirmed="true">Ireland</PLACE>
  <PLACE confirmed="true">Germany</PLACE>
  <STRATIGRAPHICPERIOD>
    <START confirmed="true">Frasnian</START>
    <END confirmed="true">Cantrabian</END>
  </STRATIGRAPHICPERIOD>
  </GEOSTRATSETS>
 </PLACES_AND_POSSIBLE_PERIODS>
 </GENUS>
```

FIGURE 6.6 XML-parsed taxonomic description.

```
<?xml version='1.0'?>

<xsl:stylesheet xmlns:xsl="http://www.w3.org/1999/XSL/Transform" version="1.0">

<xsl:template match="/">
  <HTML>
    <HEAD><TITLE>Brachiopod view</TITLE></HEAD>
    <BODY>
      <P>There are <xsl:value-of select='count( FAMILY/SUBFAMILY/GENUS )'/> genera</P>


      <P><xsl:value-of select="count( FAMILY/SUBFAMILY/GENUS[ contains( DESCRIPTION,'spine') ] )"/>
      of these contain spines.</P>

      <TABLE BORDER="2" BGCOLOR="yellow">
        <TR BGCOLOR="orange">
          <TH>Genus name</TH>
          <TH>Details</TH>
        </TR>

        <xsl:for-each select="FAMILY/SUBFAMILY/GENUS[ contains( DESCRIPTION,'spine') ]">
          <xsl:sort select='NAME'/>

          <TR>
            <TD><xsl:value-of select="NAME"/></TD>
            <TD>
              <xsl:for-each select='DESCRIPTION/DETAILS'>
                <xsl:val ue-of select="."/>;
              </xsl:for-each>
            </TD>
          </TR>

        </xsl:for-each>
      </TABLE>
    </BODY>
  </HTML>

</xsl:template>
</xsl:stylesheet>
```

FIGURE 6.7   XSL query run over XML-tagged text shown in Figure 6.6.

that would be impossible when the text is in the original stage. Over the last year, a number of computational standards have been implemented that make it much easier to run such queries over XML-coded data. Figure 6.7 shows a simple query constructed as an XSL document. When run across an XML-coded document generated from taxonomic descriptions, this query will extract a list of all those taxa in which the morphological descriptions include the term 'spine', sort them by name, and present the result as a document in a WWW browser. More complex queries would be relatively simple to program.

The XML-tagged text allows complex queries of the kind possible with highly structured databases, but in this case the problems of manually entering the data into the database have been overcome by the automatic parsing of the original text. For the example given in Figure 6.3, it would be possible to run queries that involved any possible combinations of the following:

```
- <DESCRIPTION>
  <SHELL>Shell medium to small, subcircular to elongage oval
outline, strongly biconvex </SHELL>
  <BEAK>Beak erect to sub-erect</BEAK>
  <DELTHYRIUM>delthyrium with conjunct delthyrial
plates</DELTHYRIUM >
  <ANTERIOR COMMISURE>Anterior commisure
rectimarginate
  <EXTERIOR>Exterior with faint radial ribs, and concentric growth
lamellae, each bearing small, regularly distributed, erect
spines</EXTERIOR>
  <DENTAL PLATES>Dental plates absent</DENTAL PLATES>
  <CARDINAL PROCESS>cardinal process long, thin, bifurcating
anteriorly</CARDINAL PROCESS>
</DESCRIPTION>
```

FIGURE 6.8   Showing how XML tags could be recoded to provide additional query options within morphological descriptions.

- taxon name;
- synonomy;
- author;
- date of description;
- journal or monograph;
- morphological feature or descriptive term applied to that feature;
- geological range; and
- combinations of geological age and biogeographical location.

Many advanced analytical tools developed for analysis of databases could also be applied if the tagged data were transferred to a suitable database. It would also be possible to name each of the separately tagged parts of the morphological descriptions to create within an XML document the equivalent of separate fields in databases. For example, using the first or the first two words in the hypothetical species description parsed in Figure 6.3 would generate the fields found in Figure 6.8 within the description.

This process would not always produce meaningful headings, most obviously when the name of the feature appeared not at the beginning of a phrase. However, it would allow more sophisticated queries to be run across the data and would be helpful if the ultimate aim was to use XML parsing as a step in populating biodiversity databases. During the renaming of the subsections of the description, interaction with an experienced taxonomist would allow useful names to be applied, perhaps from a glossary of appropriate terminology. An important aspect of the compilation of the brachiopod treatise was the circulation of a glossary that was agreed upon by all participants prior to the initiation of detailed taxonomic work (Curry, Connor and Simeoni 2001; Williams 2001). However, the flexibility of the procedure described means that rigid adherence to an established glossary is not essential; unusual terms will be retained whenever they are used and synonyms may be defined later.

6.7.1   ADVANTAGE OF USING XML TAGGING TO EXTRACT TAXONOMIC DATA

The major advantages of using this approach as compared to constructing a database are speed, accuracy, flexibility and ease of updating and revision. Several hundreds of samples can be processed in a short time by even a modestly powerful personal computer. XML is an extremely flexible, hierarchical language, and the parser can readily be tailored to deal with variations in the style of taxonomic writing. Updating is not a serious problem because new or revised taxonomic descriptions can simply be parsed and the resulting XML document appended to or used to replace parts of the existing resource. The text used is exactly that written is down in the monograph and has not been modified or synthesized in any way, as happens when human operators are required to extract and enter the information into database fields. This means that information is never lost; instead, interpretive views, which may be changed, are layered over it. XML is a similar language to the HTML used to construct WWW sites, and all modern browsers display XML-tagged text and the products of XSL-programmed queries, which means that the original text is always one such readily available view.

6.7.2 APPLICABILITY OF THE TECHNIQUE

There is great inherent flexibility in the XML structure to allow for differences in the structure, protocols and layouts of taxonomic descriptions in different groups. The technique could therefore be readily made applicable to the automatic XML tagging of taxonomic descriptions for a wide range of different phyla. Provided the layout and structure are standardized, taxonomic descriptions in languages other than English can also be readily parsed. Foreign language descriptions would have be accommodated in the queries run across such data, but in many cases the technical terms used to describe species are very similar in many languages.

6.8 THE FUTURE

Although adopting an entirely different approach, the technique described in this chapter is not seen as replacing databases, but rather as a complement to them. Depending on the purpose of any particular research, the XML parser could be used to rapidly extract certain particular information from a range of taxonomic descriptions. The queries can be run rapidly and reliably and with complete flexibility across data that have not been altered in any way from the original apart from the addition of XML tags.

On a wider scale, the tagged information could be transferred into an existing biodiversity database, opening up many of the advanced analytical tools that are available in such programs. Although undoubtedly speeding up the process of database population, it would be a huge task to digitally acquire the full data available from taxonomic descriptions. A more realistic strategy is for a two-stage process. First, a protocol should be set up that allows all future taxonomic descriptions to be XML tagged as it is published and, if required, incorporated into a biodiversity. The mechanisms to do this are already available and could be made virtually invisible to taxonomists, who would simply continue to use a word processor to prepare taxonomic descriptions.

This prospect suggests a range of recommendations for future developments:

- Museums and other repositories should develop mechanisms that will allow rapid dissemination of digitized taxonomic data, with suitable copyright and intellectual copyright protection.

- New publication protocols that allow synchronous printed paper publication and digital data supply should be established, again protected by suitable copyright and intellectual copyright protection as possible within the developing cyberinfrastructure developments.
- The potential of the WWW should be utilized to make existing and future taxonomic data much more widely available.
- An appreciation of the spectrum of techniques now available to structure digital data should be developed. WWW-BASED interfaces should be developed that will allow a much wider range of users and creators of taxonomic data to have access to the core resources.

ACKNOWLEDGMENTS

The authors gratefully acknowledge receipt of a BBSRC/EPSRC bioinformatics grant (BIO12052).

REFERENCES

Curry, G.B., Connor, R., and Simeoni, F. (2001) Stratigraphic distribution of brachiopods — A new method of storing and querying biodiversity information. In *Brachiopods past and present*, ed. C.H.C. Brunton, R. Cocks, and S.L. Long. Taylor & Francis, London, 424–433.

Davidson, T.D. (1887) A monograph of recent Brachiopoda, part II. *Transactions of the Linnean Society of London* 4, 75–182.

Hawksworth, D.L. (ed). 2000. The changing wildlife of Britain and Ireland. Systematics Association Special vol. 62. Taylor & Francis, London and New York. 454 pp.

Williams, A., Rowell, A.J., Muir-Wood, H., Pitrat, C., Schmidt, H., Stehli, F., Ager, D.V., Wright, A.D., Elliott, G.F., Amsden, T.W., Rudwick, M.J.S., Hatai, K., Biernat, G., McLaren, D.J., Boucot, A.J., Johnson, J.G., Staton, R.D., Grant, R.E., and Jope, H.M. (1965) Part H, Brachiopoda. In *Treatise on invertebrate paleontology*, vols. I and II. Geological Society of America and University of Kansas, 927 pp.

Williams, A., James, M.A., Emig, C.C., MacKay, S., Rhodes, M.C., Cohen, B.L., Gawthrop, A.B., Peck, L.S., Curry, G.B., Ansell, A.D., Cusack, M., Walton, D., Brunton, C.H.C., MacKinnon, D.I., and Richardson, J.R. (1997) *Brachiopoda, part H*, revised, vol. 1. Geological Society of America and University of Kansas, 1–539.

Williams, A., Brunton, C.H.C., Carlson, S.J., Alvarez, F., Blodgett, R.B., Boucot, A.J., Copper, P., Dagys, A.S., Grant, R.E., Yu-Gan, Jin, MacKinnon, D.I., Mancenido, M., Owen, E.F., Jia-Yu, Rong, Savage, N.M., and Dong-Li, Sun. (2002) *Brachiopoda, part H*, revised, vol. 4. Geological Society of America and University of Kansas, 921–1688.

7 The Grid and Biodiversity Informatics

Andrew C. Jones

CONTENTS

ABSTRACT

The Grid offers the possibility for sharing resources effectively in a distributed environment. These resources include databases that were perhaps originally intended to be used only in isolation, but which hold data that may be of broader relevance, as well as analytical tools. This chapter discusses the potential benefits of a biodiversity Grid and explains why researchers might wish to contribute to and use such a system. The notion of a Grid is introduced by describing a number of existing Grids built for other application domains. The chapter then describes how a biodiversity Grid can be designed and how it could be used, identifying problems that are particularly important to address in the biodiversity application domain. A number of Grid-related biodiversity informatics projects are described, including the BiodiversityWorld project, in which the author is one of the investigators. Finally, the chapter assesses the current state of biodiversity Grid research, identifying areas in which significant work remains to be done so that a usable, robust biodiversity Grid can be made available.

7.1 INTRODUCTION

Developments in computer technology mean that it is becoming increasingly feasible to share diverse kinds of scientific data and perform analyses on these data that would be impracticable to do by hand. For example, patterns can be searched for in a data set; possible correspondences between different kinds of data can be identified and complex hypotheses can be tested. In particular, causal links can be explored, such as links between certain genes and the incidence of a given disease, and suitable simulation software can be constructed that supports such investigations.

To date, many of the applications that have been developed to help scientists in such exploratory work have been somewhat limited in scope. For example, some are stand-alone tools that only function within a certain operating system and assume that all the data are held locally in files on the user's machine; some applications have Web 'front ends', but are fragmentary in nature — users may need to 'cut and paste' results among a number of Web applications in order to solve complex problems. Security and authentication issues are also important; we may wish to restrict access differently for each of a number of user categories and to be sure that a user is who he or she claims to be. Although there are various ways in which these problems can be solved, a uniform, standardized means of achieving flexible integration of resources in a secure environment to support scientific exploration is clearly desirable.

The computing technologies required fully to support these kinds of investigations are quite varied, but the related notions of the Grid and e-science have an important part to play. We concentrate upon the Grid and e-science and their relevance to biodiversity informatics in this chapter. We shall commence by describing the way that the Grid has developed thus far, explaining what it claims to offer and giving some classical examples of Grids. The focus of this chapter will then narrow in scope to discuss how Grids are being developed for biological science and then concentrate upon the specific needs presented by biodiversity informatics. A number of biodiversity Grid-related projects will then be described, concentrating particularly upon the BiodiversityWorld (BDW) project, of which the author is one of the investigators. It has become clear within the BDW project that the current Grid software is not entirely adequate, and we discuss next the developments that are critical if an effective biodiversity Grid is to be established. In the concluding section, we summarize how we anticipate a biodiversity Grid developing to the point where it can be used routinely by researchers working in the biological sciences. It is our aim here to explain how a biodiversity Grid might transform the way in which computers are used in biodiversity research, as well as to inform researchers in the field of biodiversity informatics about software developments needed in order to make such a Grid fully implementable. It should be noted that this chapter was written in early 2004, so a number of recent developments (for example in relation to Globus and SRB software) are not therefore covered.

7.2 THE GRID AND E-SCIENCE

The Grid was originally conceived as a means by which large amounts of computational power could be made available to individuals on demand in a readily usable way. An analogy sometimes given is that of national electricity grids: plugging in an appliance, one has immediate use of mains electricity, with no need to specify which power station generated the electricity being used. Similarly, proponents of the Grid want to supply a computing infrastructure where many of the mundane aspects of how a particular task is to be organized can be hidden from a user.

Foster and Kesselman (1999) define a computational Grid as a hardware and software infrastructure that provides dependable, consistent, pervasive, and inexpensive access to high-end computational facilities'. This kind of definition implies that high-performance computing (HPC) applications are particularly intended to benefit from the Grid. We shall see later that it is certainly true that HPC applications can make good use of Grid software, but there are also other reasons why a computational Grid is desirable — for example, to provide a uniform software architecture in which resources (data sources, software) can be located and their use coordinated in order to perform some complex, collaborative, but not necessarily computationally expensive task. This shift in emphasis is discernable in Hey and Trefethen's (2002) more recent definition:

> e-Science is about global collaboration in key areas of science and the next generation of infrastructure that will enable it. The infrastructure to enable this science revolution is generally referred to as the Grid.

It should be understood that although the term *Grid* is sometimes used in connection with specific software (e.g., Globus, described later), it also denotes a concept — that is, any infrastructure that can fully support e-science. It is particularly the latter sense of the word that is of interest, especially since arguably the full infrastructure is not yet in place, and the details have not yet been fully worked out, as we shall discuss later.

In the remainder of this section we shall describe some of the important characteristics of Grid infrastructures that have been developed thus far and provide some classical examples of Grids.

7.2.1 Grid Infrastructures

The Grid is often associated very closely with the Globus software, but a variety of developments is relevant to realization of the concept of the Grid. One important aspect for HPC applications is the development of appropriate hardware infrastructure. The TeraGrid project [1] is an example of this. But it is not necessarily the case that all Grid users will need to have direct access to a high-performance network. On the other hand, the development of standard appropriate Grid software is essential if the Grid is to become widely used. Note that there is nothing to prevent Web-based user interfaces being built to Grid-based applications if specialized interfaces are not required. In this case, the user does not have to install any special software to use applications that, somewhere within their architecture, exploit Grid concepts and software. We can enumerate a number of categories of Grid software, including data storage, scheduling and general Grid middleware.

Grid data storage software includes:

- Storage Resource Broker (SRB) [2], which provides facilities for what can be regarded as a distributed file system; and
- Open Grid Services Architecture — Database Access and Integration (OGSA-DAI) [3], which builds upon OGSA (see following text), providing database access and distributed queries in a Grid environment.

SRB is functionally fairly limited, but is useful for situations in which the main technical challenge is to provide distributed access to large data sets, using replication and other techniques in order to improve access speed. OGSA-DAI is much more appropriate if one is typically wanting to access small parts of a data set, but it should be noted that the OGSA-DAI software implementations are still at a prototypical stage.

An example of scheduling software is CONDOR [4]. This software distributes applications according to the load upon a network of machines, in order to make good use of the available computational power.

Grid middleware aims to provide a more complete set of facilities for the development of Grid applications. Major Grid middleware examples include:

- Globus [5] is perhaps the most widely known Grid middleware software. It provides mechanisms for controlling software in a distributed, secure environment. We shall consider Globus in more detail below.
- UNICORE [6] is an alternative to Globus that has been developed particularly to support workflows comprising tasks to be performed in a distributed environment. It lacks some features of Globus, such as support for multiple programming languages.
- Legion [7]. This aims to provide a virtual operating system for distributed resources (i.e., to make it possible in some sense to view an entire network of computers as if they combine to form a single, powerful distributed computer).

These Grid middleware alternatives differ quite substantially from each other, and each is only a partial solution to the problem of providing middleware to enable Grid applications to be developed readily. Indeed, some researchers are seeking to exploit the features of more than one product simultaneously (e.g., in the Interoperability Project (GRip) [8] project, whose aim is to develop software to enable interoperability between UNICORE and Globus).

Since the Globus project is perhaps the most widely known, we shall now discuss how Globus has evolved so that we can illustrate the strengths and limitations of a typical example of Grid software at present.

Early versions of Globus concentrated particularly on software and protocols addressing a fundamental concern: how tasks could be executed efficiently on machines at remote locations and how data could be transferred without compromising the security of these remote machines. Globus Toolkit 2 provides a number of facilities that address these concerns, including the Globus security infrastructure (GSI), which uses a *gatekeeper* to validate users' credentials and provides well-defined restricted access to remote resources; the Globus access for secondary storage (GASS) facility, which provides data transfer facilities; and the Globus resources allocation and management (GRAM) service, which allows remote invocation of tasks. One additional important feature is the metacomputing directory service (MDS), which provides some basic support for storing descriptions of facilities offered by various resources, information about their location, etc.

The Globus facilities we have just summarized make it possible to build applications that need large amounts of computational power — for example, in order to execute time-consuming algorithms — but provide only weak support for more knowledge-intensive tasks. For example, we might wish to discover all resources that hold information on a particular topic, or we may wish to synthesize information from a set of knowledge sources that represent their information in quite distinct ways. A significant step forward has been made in Globus Toolkit 3, in that grid services are now supported, via the open grid services architecture (OGSA) [9]. These services are very similar to Web services, and the possibility is opened up of using Web service-related facilities, such as UDDI (universal description, discovery and integration of Web services), to standardize the way that resources are described and located. Another important consequence of this development is that one is now much less restricted to a particular platform, such as UNIX, because the services are advertised and used in a platform-independent manner. This makes interoperability — the ability to use resources together even though they may be heterogeneous (i.e., they may differ) in various senses — much easier to achieve than with the earlier Globus architecture.

But it must be understood that there is still a great deal of room for further development. For example, OGSA is not entirely compatible with ordinary Web services; a recent announcement (WSRF, the Web service resource framework [10]) seeks to address this issue, but at the expense of further instability in Globus specifications. Although such service-orientated computing is an attractive element of a Grid, research and development is still under way; for example, extensions of UDDI are being proposed (e.g., ShaikhAli et al. 2003). Furthermore, service-orientated computing is not all that is needed. For example, standard tools for visualization, for building ontologies (which hold definitions of terms such as *taxon* or *character* and their relationships to each other), etc. must be developed if the Grid is to provide effective support for e-science. We shall discuss this matter in more detail at the end of the chapter.

7.2.2 CLASSICAL EXAMPLES OF GRIDS

The most long-established Grids are mostly designed to allow complex computation, simulation and/or sharing of large data sets. Many of these data sets are relatively simple in structure (e.g., some are flat files); the main problem they present is often their very large size. We shall look briefly at three examples chosen primarily to illustrate that the needs of biodiversity informatics, which we shall consider later, are somewhat different from those of more mainstream Grid applications and place different requirements upon the Grid infrastructure that should be built.

The *GriPhyN* (Grid Physics Network [11]) project focuses on the creation of a Grid to support the storage of data derived from experiments or simulations in high-energy physics and the provision of mechanisms for efficient delivery of data, on demand, to researchers. Among other things, this involves the use of techniques for distributed replication and caching of data and the development of mechanisms for ensuring that quality of service criteria (requirements for performance, cost, reliability) are met.

The *Earth System Grid* [12] is a distributed system that aims to store and provide access to the large amounts of data generated by climate simulations, as well as to enable users to discover data relevant to their interests (perhaps models related to the one currently under consideration, for example).

The *National Virtual Observatory* (NVO) [13] provides a Grid in which very large, distributed data sets generated as a result of astronomical observations are stored and can be analysed (e.g., to discover rare objects). Such analyses are only practicable because of the large numbers of observations that can be mined automatically in a system like this.

For more information about these and other projects, see Johnston (2002). Grids such as these rely primarily on high-performance computing and data storage facilities to support scientific exploration. We shall now turn to the narrower field of bioinformatics and discuss the ways in which Grids can be and are being developed to support research in that domain, including efforts to increase the emphasis on using explicit knowledge, rather than merely executing analytical algorithms.

7.3   THE GRID AND BIOLOGICAL SCIENCE

There is a variety of Grid-based projects relevant to biological science. Some of these are addressing fairly specific problems (e.g., modelling some specific biological process), while a smaller number are seeking to build a generally useful bioinformatics Grid.

An example of a Grid designed to address a specific problem is BioSimGrid [14], which is designed to make the results of biomolecular simulations performed in various laboratories available so that comparisons can be made of results obtained using differing algorithms and data. Another project seeking to achieve similar aims is Bio-Grid [15]. One of the most important differences between BioSimGrid and Bio-Grid, from a computing point of view, is that the latter is part of a larger Grid project, EuroGrid, which is seeking to build a European grid infrastructure that is of general use — not just for one application. The role of Bio-Grid within this larger project is as one of the exemplars chosen to test the Grid they are developing.

The focus for EuroGrid is, as for many Grid projects, upon HPC systems. One important project that is looking at a broader range of Grid requirements for bioinformatics is myGrid [16]. This is a large project, with a strong computer science emphasis. Some of the specific issues being addressed are the development of an ontology to help classify and support discovery of resources; the storage of provenance data, to describe, for example, derivation of a result; and the capture and enactment of scientific workflows. Thus, myGrid is not focused only upon infrastructures for efficient algorithm execution, but also on the use of more descriptive data. The kinds of tasks supported include helping in the accumulation of evidence and the pursuit of collaborative research in a customizable environment where users can specify preferences (e.g., to guide selection of resources of interest to them). From what we have been able to determine, the emphasis is particularly upon the infrastructure required to provide these facilities, although some biological exemplars are being developed within the project.

An ultimate goal for biological science might be to build a Grid in which very disparate data, from molecular through to information about groups of organisms (e.g., species distribution), were made available in such a way that new hypotheses could be expressed, which can only be tested by reasoning with data of such diverse kinds simultaneously. This vision is indeed one of the foci of the life sciences grid (LSG) research group of the Global Grid Forum [17], but its full realization is still well in the future. As we turn now to biodiversity informatics, it is clear that, to some extent, the same need applies there, but with extension to other, abiotic sources of information too.

7.4   THE GRID AND BIODIVERSITY

Biodiversity informatics databases and tools have some particular characteristics that make the construction of a biodiversity Grid an ambitious aim. We shall first discuss these characteristics and then discuss, from a user's point of view, what capabilities a biodiversity Grid might offer — how one might wish to be able to use existing and new resources within a Grid-based environment. We shall then outline a number of Grid-related biodiversity informatics projects, after which we shall provide some detail of how the BiodiversityWorld project (in which the author is an investigator) is developing as a test bed in which ways of meeting some of the requirements identified can be explored. This, as far as we are aware, is the only current project that focuses specifically upon building a biodiversity Grid.

7.4.1 Biodiversity-Related Databases and Tools

A wide range of differing kinds of data has been collected by scientists involved in biodiversity research. For example, specimen data may include information about where the specimen was collected, the collector, morphological details, etc. Species data may be primarily descriptive, or it may be primarily taxonomic in nature (e.g., a taxonomic checklist). Other data will also be relevant (e.g., gene sequence, geographical and climate data, depending on the scientist's interests), as will derived data (e.g., sets of phylogenetic trees generated by an appropriate algorithm). These data have a number of characteristics that make their management and use in a Grid potentially difficult, including the following:

- A wide range of different kinds of data is needed in order to perform some tasks of interest, such as bioclimatic modelling; all these data need to be made available within a Grid environment.
- The databases may be heterogeneous in various ways. Different database management systems may have been used; similar data may not be represented uniformly in all databases (there may be variation in the structure of the data, or in the terms and units used or in both). This latter kind of variation is an example of *semantic heterogeneity*.
- Furthermore, a database may often have been built for a very specific purpose, and this may have influenced its design, both in terms of what information is stored and its organization. In contrast, many important data sets in other disciplines, including bioinformatics, are generally stored in agreed formats in public repositories.
- Some data are sensitive (an obvious example being data pertaining to endangered species), so precise and dependable access control is required to ensure that users see only the data they are entitled to have access to.
- The data may vary in ways relating to scientific opinion. For example, not all data pertaining to a given species may be stored in association with the same scientific name if there are differences of taxonomic opinion.
- Data quality issues of a number of different kinds arise. For example, a scientist may incorrectly identify a specimen and record his or her observations under an incorrect scientific name, or he or she may forget to include a negative sign in latitude/longitude coordinates for the location where a specimen was collected.

These issues mean that existing Grid middleware is insufficient. Tools to deal with semantic heterogeneity, data quality problems and differing professional opinion are needed, to name only a few. Similarly, when we come to biodiversity-related tools, we encounter some distinctive characteristics, including the following:

- Some tools were written to solve very specific problems and they assume data input and output in proprietary formats.

- Many tools were written with the intention that they be used as stand-alone applications running on a single platform, such as an Apple Macintosh, in contrast with mainstream bioinformatics tools, which are often open source or accessible via a Web user interface or as a Web service.

If such tools are to be integrated into a Grid environment, they need to be re-engineered, specialized wrappers need to be created to drive the applications as if they were a user, or the tools need to be used via their existing user interface on the user's machine or via remote control software such as VNC [18] outside the Grid.

7.4.2 How Might a Biodiversity Grid Be Used?

The author's vision of a fully established biodiversity Grid is of a computing infrastructure that would fulfill some important roles:

- It would act as an information source, by means of which users could retrieve information on some particular topic from a wide range of distributed databases through a single user interface that provides an integrated view of the Grid. This is in contrast with the current situation in which a user may well need to visit a number of Web sites, including a search engine, before he or she can locate and assemble the required information.
- It would allow the user to search for resources having particular characteristics — for example, a resource that holds data on the geographical distribution of a particular family of plants, or a resource that generates phylogenetic trees (i.e., resource discovery).
- It would allow the user to assemble a complex workflow in which analytical tools and information from various resources can be combined to perform a complex analysis requiring a number of distinct stages, using quite varied kinds of data.
- It would allow the user to access results generated from laboratory equipment connected to the Grid and provide a means — perhaps using a Tablet PC or similar device — of recording observations and experimental provenance data. This would make it easier for experiments to be repeated. The *in silico* parts could be rerun, perhaps modifying some aspects (e.g., using a different algorithm or changing the scientific method represented in the workflow); experimental data previously captured could be re-used, or new experiments could be performed and passed through the same analytical process as the previous experimental data, etc.
- It would provide sophisticated knowledge-based tools to help users formulate and express scientific hypotheses — perhaps even using natural language (such as English)? — and design experiments to test such hypotheses.
- It would provide a user interface allowing the user to access Grid resources in a seamless manner. Differences of communication protocols, data representation, etc. would be managed in such a way that the user would not necessarily need to be aware of them.

We shall discuss how this might be achieved in a later section. Suffice it to say that, at present, some of these requirements need further research before they can be implemented in a usable form. Currently, efforts to implement a biodiversity Grid are fragmentary, as we shall now see.

7.4.3 SOME GRID-RELATED BIODIVERSITY INFORMATICS PROJECTS

In this section we shall describe briefly three Grid-related biodiversity projects, none of which professes to build a full Grid, but each of which has some significant contribution to make.

The GRAB and Grid and Biodiversity project (Jones et al. 2002) was, as far as the author is aware, the first attempt to use Grid technology to build a prototype biodiversity Grid. In the project, the author and others were asked to explore the use of Globus and SRB within biodiversity informatics, and a small prototype was built that used the SPICE for Species 2000 catalogue of life [19], two species databases (ILDIS [20] and FishBase [21]) and a database built from public domain climate information obtained from the U.S. National Climate Data Center (NCDC) [22]. The catalogue of life is used to determine the scientific names needed to query the species databases; crude climate envelopes can be created, based on the species-related information retrieved, and species native to geographical regions falling within a given climate envelope can be retrieved.

We found that SRB offered no features that were relevant to our project and that, in the state it was in at the time, Globus offered only limited benefits. In fact, the process by which the GRAB system was built gave us some useful insight into the relevance of Globus. We first made the databases available in a non-Grid environment using HTTP and XML and then replaced this software interface by one using Globus. This proved difficult to achieve because the Globus mechanisms for transferring data and executing remote jobs did not relate in a natural way to the request/response model that we initially implemented. Some benefit was obtained from use of the Globus MDS as a catalogue of resources, but overall our conclusion was that the benefits offered by Globus at the time were very limited. This situation is now changing, as we have already seen.

The WhyWhere project [23] makes it possible to prepare a very large environmental database and to search for correlations with location data (e.g., in order to support the study of migration patterns). Grid technologies — specifically SRB — are used in order to provide high-performance computing facilities that process ecological and other data, but this is a hybrid system in which another mechanism, the Species Analyst [24], is used to access the location data. Thus, the Grid is used for a part of the system that particularly benefits from its use, in preference to the construction of a full bioclimatic modelling Grid.

The SEEK project [25] is addressing some important issues of relevance to biodiversity- and eco-informatics, including the notions of semantic mediation (techniques for supporting variations in the way that data are represented and should be interpreted) and ecological ontologies (ecological terms and relationships between them). It would appear that the Grid is a secondary concern, although this project is working in areas of direct relevance to the Grid. The intention of SEEK is to make it easier for researchers to gain global access to ecological information and use distributed computing services in an environment that supports the construction and use of explicit analytical processes (workflows).

Each of these projects has features that would be beneficial in a full biodiversity Grid as individual techniques or software that could be beneficially incorporated directly into a Grid environment. We now turn to the BiodiversityWorld project, which aims to take a less fragmentary approach to building a biodiversity Grid.

7.4.4   THE BIODIVERSITYWORLD PROJECT

BiodiversityWorld (BDW) (Jones et al. 2003; White et al. 2003) is a three-year e-science project, funded by the UK BBSRC, in which we are exploring how a problem-solving environment (PSE) can be designed and developed for biodiversity informatics in a Grid environment. It is motivated by three exemplars in the field of biodiversity informatics, but we are concerned to make this system extensible to be of general use within the discipline. Our aim is to provide scientists with tools with which they can readily access resources that may have originally been designed for use in isolation, composing these resources into complex workflows and making it as straightforward as we can for new resources to be created and introduced to the system. The three chosen exemplars are bioclimatic modelling and climate change, biodiversity richness and conservation evaluation and phylogeny and palaeoclimate modelling:

- In the *bioclimatic modelling and climate change* exemplar, a bioclimatic model describing the envelope of climate and ecological conditions under which a single species lives is generated. This model can then be used for various kinds of predictions, such as the future distribution of the species under changing climate conditions. In the BDW environment it is possible to rerun such analyses many times, changing the species, the analysis tools used, etc. as desired.
- In the *biodiversity richness analysis and conservation evaluation* exemplar, distribution data are used with appropriate metrics to assess biodiversity richness. The results of this kind of analysis could potentially be used to inform conservation policy.
- In the *phylogeny and palaeoclimate modelling* exemplar, phylogenies are to be used to interpret other biodiversity data, such as distribution. This will allow a variety of biological questions to be addressed (e.g., is geography a good predictor of relationship among species lineages? Have lineages stayed put, adapting *in situ* while climates have changed?)

Each of these exemplars requires a wide range of diverse kinds of data, many of them having their own complex structures, such as climate data, species distribution data and gene sequence data. Also, a common theme in the workflows we have thus far envisaged is taxonomic verification, in which scientific names are checked against a taxonomic catalogue (provided at present by the SPICE for Species 2000 catalogue of life) to determine the names that might be used in searching these other data sources. Similarly, a range of analytic tools is required, each with its implementational idiosyncrasies.

As a Grid-based project, BiodiversityWorld has distinctive characteristics. High-performance computing resources are only of use for a small range of self-contained tasks in biodiversity informatics; the wide range of types of data to be used requires careful management. On the other hand, a limited range of operations on these data sources is typically required; for example, the SPICE system provides six operations, of which two (search for scientific name and retrieve species information) suffice for the tasks we currently envisage. Thus, it does not seem necessary to provide a general query facility (e.g., SQL); a set of services defined for each resource is sufficient. To make access to these data sources and tools possible from BiodiversityWorld, a generic access mechanism must be achieved while retaining flexibility as to the operations to be performed by each resource and the data types they can use. These heterogeneous resources must therefore be wrapped, and metadata must be available to indicate how they are to be used.

A further consideration influencing the design of the BDW architecture is that, typically, in the scenarios we envisage, an individual scientist will not be collaborating interactively with other users of the system. This and the kinds of operations required for each resource imply that a service-orientated architecture is appropriate. However, an important additional requirement is that users should be able to manipulate certain kinds of data interactively (e.g., for selection among a set of trees generated as a result of phylogenetic analysis). But whatever software architecture is adopted, it should not be bound too closely to a particular Grid infrastructure. Grid software is rapidly evolving — for example, there are major differences between Globus Toolkit versions 2 and 3 — and it is desirable, as far as possible, for migration to a new infrastructure to have minimal impact on the resources. An interoperation framework that is not tied to a specific Grid infrastructure has therefore been designed.

The architecture we have adopted is illustrated in Figure 7.1. A layer of abstraction is placed between BiodiversityWorld (BDW) components and the Grid, which we refer to as the BiodiversityWorld–Grid interface (BGI). This means that if the Grid infrastructure changes, only the BGI needs to be re-implemented. It is intended that other components will remain unchanged. In order to insulate the BiodiversityWorld system from the resources' heterogeneity, we wrap these resources and provide an invocation mechanism that allows any operation to be invoked in a standard manner. The available operations are specified by metadata associated with these resources. Furthermore, to insulate these wrappers from changing Grid infrastructure, they need to be wrapped by wrappers providing an interface to the Grid infrastructure. This further wrapping is provided by the BGI layer illustrated in Figure 7.1. To deal with the interactivity mentioned previously, there are a number of possibilities; the one that we currently support is simply for all of this processing to be carried out on the user's local machine.

We have thus far implemented two versions of the BGI: one uses Java RMI (a technology, unrelated to the Grid, illustrating the general nature of our architecture); the other uses grid services (OGSA), as provided in Globus Toolkit version 3.

The BiodiversityWorld project is a serious attempt to develop a biodiversity Grid but, as in any project, we have had to make some simplifying decisions. A major simplification has been to assume that any high-performance computing requirements will be carried out by a single logical BDW resource. This resource might be implemented as a cluster of machines or it might use Grid middleware, but anything of this nature is hidden away behind its single point of presence on the BDW Grid. Another simplification is that a full ontology is not likely to be created within the lifetime of the project, but rather only a thesaurus of related terms. To some extent, this limits the effectiveness of resource discovery tools since, if a resource is described using different terms from the user's search terms, it can only be found if there is a one to one synonymic correspondence between the terms provided by the user and those used within the system. Also, reasoning with differing data representations is limited by this decision, for example. But our main concern has been to build a complete system that can be used by biologists to assist them in their research and hence to demonstrate the potential benefits of a biodiversity Grid.

```
<?xml version='1.0?>

<xsl:stylesheet xmlns:xsl="http://www.w3.org/1999/XSL/Transform" version="1.0">

<xsl:template match="/">
  <HTML>
    <HEAD><TITLE>Brachiopod view</TITLE></HEAD>
    <BODY>
      <P>There are <xsl:value-of select='count( FAMILY/SUBFAMILY/GENUS )'/> genera</P>


      <P><xsl:value-of select="count( FAMILY/SUBFAMILY/GENUS[ contains( DESCRIPTION,'spine') ] )"/>
      of these contain spines.</P>

      <TABLE BORDER="2" BGCOLOR="yellow">
        <TR BGCOLOR="orange">
          <TH>Genus name</TH>
          <TH>Details</TH>
        </TR>

        <xsl:for-each select="FAMILY/SUB FAMILY/GENUS[ contains( DESCRIPTION,'spine') ]">
          <xsl:sort select='NAME'/>

          <TR>
            <TD><xsl:value-of select="NAME"/></TD>
            <TD>
              <xsl:for-each select='DESCRIPTION/DETAILS'>
                <xsl:value-of select="."/>;
              </xsl:for-each>
            </TD>
          </TR>

        </xsl:for-each>
      </TABLE>
    </BODY>
  </HTML>

</xsl:template>
</xsl:stylesheet>
```

FIGURE 7.1

7.5   TOWARDS A BIODIVERSITY GRID

In this chapter we have described the Grid and presented some examples of Grid applications in various disciplines, including biodiversity. The BiodiversityWorld project, though aiming to build a prototype biodiversity Grid, will inevitably fall short of achieving everything necessary to this end within the fixed time period for which it has been funded. Therefore, here are a number of developments necessary for a biodiversity Grid to be fully implemented:

- a *biodiversity ontology* to support sophisticated resource discovery, workflow construction, etc.;
- *tools* for *knowledge discovery* (such as have been prototyped in the knowledge Grid; see Cannataro and Talia, 2003);

- *inter-Grid interoperability* because, at present, most Grids are built with a view to enabling interoperability only within one specific Grid. BDW provides a partial solution to this problem. It can interoperate with resources that conform to or are wrapped to conform to the BGI API, and it is not dependent on specific Grid middleware because the BGI is designed to be built on top of the Grid middleware of one's choice. But this assumes that BDW is to subsume other Grids, which is not realistic, especially given the design assumption underlying BDW that high throughput between BDW resources is not required;
- ideally, a *standard Grid client*, analogous to a Web browser, that supports common tasks (workflow construction, resource discovery, etc.) and also has a plug-in architecture so it can be extended as necessary for individual Grid applications;
- *stable standards*, including standards for how Grid services are described and used, and how workflows are specified;
- conventions and mechanisms for sharing data in *virtual organizations* (Foster et al. 2001; Goble and De Roure 2002) so that fine control can be exercised over access rights (e.g., an individual might be given access to sensitive data, but only for the duration of a particular collaborative project);
- *tools* for Grid *resource development* and for Grid *maintenance* (e.g., for registering new resources); and
- a *critical mass* of biodiversity resources (databases and tools) that is great enough for a wide range of biodiversity-related tasks to be performed conveniently on the Grid.

7.6 SUMMARY AND CONCLUSIONS

We have seen that Grid computing is an area in which interesting developments have been made, but that further progress is needed before it becomes something that will be routinely used in scientific research. The Grid is a natural unifying technology for biodiversity informatics, where many small data sets exist, distributed across the world, that could be brought to bear on a range of important problems. But this technology needs to be stabilized and well documented if such a Grid is to be effective. Biodiversity informatics has characteristics, detailed earlier, that mean it is essential for research into biodiversity-specific Grids to continue in BiodiversityWorld and, hopefully, in other projects so that the Grid community can be informed of these characteristics and of how they can be accommodated in the Grid.

ACKNOWLEDGMENTS

The BiodiversityWorld project is funded by a research grant from the UK Biotechnology and Biological Sciences Research Council (BBSRC); the GRAB project was funded by a grant from the UK Department of Trade and Industry. Although the views expressed in this chapter are solely the author's, their formation has been influenced by interaction with colleagues within the BiodiversityWorld team — especially with Prof. W.A. Gray and Dr R.J. White — and also with Dr Peter Kille of Cardiff University.

REFERENCES

Cannataro, M. and Talia, D. (2003) The knowledge grid. *Communications of the ACM* 46(1): 89–93.

Foster, I. and Kesselman, C., eds. (1999) *The Grid: Blueprint for a new computing infrastructure.* Morgan Kaufmann, San Francisco, CA.

Foster, I., Kesselman, C., and Tuecke, S. (2001) The anatomy of the Grid: Enabling scalable virtual organizations. *International Journal of Supercomputer Applications* 15(3): 200–222.

Goble, C. and De Roure, D. (2002) The Grid: An application of the semantic Web. *ACM SIGMOD Record* 31(4): 65–70.

Hey, T. and Trefethen, A.E. (2002) The UK e-Science Core Programme and the Grid (http://www.rcuk.ac.uk/escience/documents/rreport_CoreProgGrid.pdf).

Johnston, W.E. (2002) The computing and data grid approach: Infrastructure for distributed science applications. *Journal of Computing and Informatics* 21(4): 293–319.

Jones, A.C., Gray, W.A., Giddy, J.P., and Fiddian, N.J. (2002) Linking heterogeneous biodiversity information systems on the Grid: The GRAB prototype. *Journal of Computing and Informatics* 21(4): 383–398.

Jones, A.C., White, R.J., Pittas, N., Gray, W.A., Sutton, T., Xu, X., Bromley, O., Caithness, N., Bisby, F.A., Fiddian, N.J., Scoble, M., Culham, A., and Williams, P. (2003) BiodiversityWorld: An architecture for an extensible virtual laboratory for analysing biodiversity patterns. In *Proceedings of the UK e-science all hands meeting.* EPSRC, Nottingham, UK, 759–765.

ShaikhAli, A., Rana, O.F., Al-Ali, R., and Walker, D.W. (2003) UDDIe: An extended registry for Web services. In *Proceedings of the UK e-science all hands meeting.* EPSRC, Nottingham, UK, 731–735.

White, R.J., Bisby, F.A., Caithness, N., Sutton, T., Brewer, P., Williams, P., Culham, A., Scoble, M., Jones, A.C., Gray, W.A., Fiddian, N.J., Pittas, N., Xu, X., Bromley, O., and Valdes, P. (2003) The BiodiversityWorld environment as an extensible virtual laboratory for analysing biodiversity patterns. In *Proceedings of the UK e-science all hands meeting.* EPSRC, Nottingham, UK, 341–344.

CITED WWW RESOURCES

1. http://www.teragrid.org/
2. http://www.sdc.edu/srb/index.php
3. http://www.ogsadai.org.uk/
4. http://www.cs.wisc.edu/condor/
5. http://www.globus.org/
6. http://www.unicore.org/
7. http://legion.virginia.edu/
8. http://www.grid-interoperability.org/
9. http://www.globus.org/ogsa/
10. http://www.globus.org/wsrf
11. http://www.griphyn.org/index.php
12. http://www.earthsystemgrid.org/
13. http://www.us-vo.org/
14. http://www.biosimgrid.org/
15. http://biogrid.icm.edu.pl/
16. http://www.mygrid.org.uk
17. http://www.ggf.org/7_APM/LSG.htm
18. http://www.realvnc.com/
19. http://www.sp2000.org/

20. http://www.ildis.org/
21. http://www.fishbase.org/home.htm
22. http://www.ncdc.noaa.gov/oa/ncdc.html
23. http://biodi.sdsc.edu/ww_home.html
24. http://speciesanalyst.net/
25. http://seek.ecoinformatics.org/

8 LIAS — An Interactive Database System for Structured Descriptive Data of Ascomycetes

Dagmar Triebel, Derek Peršoh, Thomas H. Nash III, Luciana Zedda and Gerhard Rambold

CONTENTS

ABSTRACT

LIAS is a multi-authored database system for descriptive and related biodiversity data on lichens and non-lichenized ascomycetes. In 2004 it contained about 5500 species-level and 850 genus-level records. Various Web interfaces are provided for editing and querying the data. Aside from this major goal, LIAS has meanwhile gained importance with respect to (1) the general demand for rapid identification of organisms; (2) the demand for geospatial distribution of organisms; and (3) the demand for name pools. For enabling coverage of these aspects, three subprojects — LIAS light, LIAS checklists and LIAS names — were set up.

8.1   INTRODUCTION

The development of storage and retrieval systems for biodiversity data is considered a central task under the aspect of sustainable data provision for future research activities in the context of various international biodiversity initiatives and programmes such as GBIF (Global Biodiversity Information Facility), GTI (Global Taxonomy Initiative), Species 2000 (indexing the world's known species) and DIVERSITAS (International Global Environmental Change Research Programme). The major goal of these is to gain and provide new scientific insights and to prepare a profound data basis for subsequent national and regional research initiatives. For such biodiversity long-term projects, however, access is required to structured data that do not concern spatial information only. For instance, data concerning ecology, morphology and chemistry of the respective organism groups are of interest as well. Therefore, these projects are in strong need of databases for the storage and management of a broad spectrum of specimen- and taxon-related information.

Furthermore, tools for data analysis (e.g., for determining the inter-relation of taxonomic information with biotic and abiotic environmental parameters) are required. Information of this kind needs to be stored and maintained in a range of more or less specialized databases. While a highly modularized database suite, like Diversity Workbench WWW or less modularized systems like Systax (http://www.biologie.uni-ulm.de/systax/) and Specify (http://software.org/specify) are designed for the storage of taxonomic information, it is the domain of other systems like GIS to contain climatic, edaphic and ecological data of different kinds. Speciesbank initiatives collecting and providing descriptive data are currently not covered by any of the work programme areas of the current GBIF working plan (Anonymous, GBIF Work Programme 2004). Nevertheless, many different projects are already attempting species-level synthesis of data from multiple sources and it is considered to be important by GBIF International to investigate the possible options and models for that purpose.

LIAS (http://www.lias.net), 'a global information system for lichenized and non-lichenized ascomycetes', is an example of a multi-authored database system containing descriptive and structurally similar biodiversity data on lichens and non-lichenized ascomycetes. The system is based on Diversity Workbench database components and is a user-oriented online service for establishing structured descriptive data for ascomycetes. A central task in this context is the development of interactive Web tools to allow scientists' easy access and user-friendly individual remote editing of the data. The online database system is offered for multiple usage and thus dissemination of expert knowledge (while respecting intellectual property rights of data contributors), mainly by providing public access to up to date interactive identification keys and database-generated natural language descriptions of ascomycetes. In addition, it promotes common standards for descriptive data connected with taxonomic names of ascomycetes to facilitate databank interoperability and data exchange.

8.2   FROM 1993 UNTIL 2004

In 1993, the project was initiated at the Botanische Staatssammlung, Munich, under the programmatic title 'information and data storage system for lichenized and lichenicolous ascomycetes' (LIAS). The basis of the data was a collection of descriptive data of lichen genera coded in the DELTA format. Two years later, a set of HTML pages with information on the LIAS project was put on the Internet. The DELTA data collection grew due to continuous descriptive species and genus-level data entry. In 1996, a first set of LIAS key modules was ready for download and local usage with Intkey (Rambold 1997). For compiling from DELTA data binary code to be used for the interactive key application Intkey, the compiler software Confor (being part of the CSIRO DELTA package) was applied (Dallwitz et al. 1993 onwards). At that time, the key module, 'genera of lichenized and lichenicolous ascomycetes', represented the core element of the whole LIAS system (Rambold and Triebel 1995 to 2004), and a number of species-level data sets created by various lichen experts became available as well.

In 1997, an important technical step forward was achieved when genus data were transferred in the relational MS Access database DeltaAccess (now DiversityDescriptions), combining the advantages of DELTA and a relational database system (Hagedorn 2001a). For generating HTML data entry forms for browser-based data input, specific report functions were implemented (Hagedorn and Rambold 2000). Apart from the two Web interfaces DeltaAccess Perl Script (DAP) and DeltaAccess Web Interface (DAWI) (both as beta versions), taxon subset-specific initialization files on the Web server were installed for usage with Intkey as an auxiliary application. In this context, the key modules for download and local usage were abandoned. Rambold and Hagedorn (1998) first published a scientific evaluation of diagnostic characters based on the LIAS generic data set.

Also at that time, the sets of species-level data were continuously growing and the number of lichenologists and mycologists using the LIAS service increased by starting their own family subprojects. In 1999, the concept of LIAS was thematically extended towards the inclusion of non-lichenized and non-lichenicolous ascomycetes; as a consequence, the project subtitle was changed. Within that same year, all species data hitherto collected were imported into an extended version of DeltaAccess (as part of the Diversity Workbench now under the name DiversityDescriptions). At that time, the LIAS main databank included 614 mostly multistate characters. For each family project, its own subproject with a selected number of characters and states was generated and dynamically linked to the main database. At the end of 2000, LIAS contained more than 2700 data sets with descriptive characters of ascomycete taxa, 1900 of which were data sets at species level. The LIAS main database was enlarged to have 723 mostly multistate characters. From that time, LIAS was presented under its own domain name: www.lias.net.

During 2000 several technical and thematic cooperations between LIAS and the following online projects were initiated: the Global Information System for the Biodiversity of Plant Pathogenic Fungi (GLOPP, www.glopp.net), MYCONET (see Eriksson et al. 2004) and the Lichen Checklists Project. A set of more than 600 records with descriptive data of Erysiphalean fungi resulted from the initial cooperation (Triebel et al. 2003a), while 14,135 databased checklist records of lichens and lichenicolous fungi arose from the latter (see http://checklists.lias.net/). One year later, the LIAS Descriptors Workbench was started as collaboration between Arizona State University and the LIAS project team. As a result the LIAS glossary Wiki is available now under http://glossar.lias.net.

8.3   THE STATUS QUO OF TECHNOLOGY IN 2004 AND THE UNDERLYING DATABASE APPLICATION DIVERSITYDESCRIPTIONS

DiversityDescriptions is a free database application in MS Access for Windows that builds up a relational structure from DELTA-based data during import. In 2002, the LIAS main database was transferred from DeltaAccess to an advanced version of DiversityDescriptions exhibiting various new specifications. Remote access for correcting and adding data was enabled by database-generated HTML data entry and revision forms to be submitted to the server for data update via CGI. The HTML data forms were adopted separately for each family so that each family set of data contained only characters and character states meaningful in the specific context. In addition to HTML form views, natural language descriptions in HTML were provided for better human reading and to be used as templates in monographs and floras, such as the 'lichen flora of the Greater Sonoran Desert region' (Nash et al. 2002a, b).

DiversityDescriptions is a component of the Diversity Workbench database suite (Figure 8.1). The Diversity Workbench modularizes the information models used in biodiversity research and creates a framework of exchangeable components, each of which is specialized for an information area. The components interact through minimized interface definitions and without knowledge of the internal operation of each other. Major components are DiversityCollection for georeferenced specimen collection data, DiversityReferences for literature data and DiversityDescriptions for descriptive data. The information models are published (see Hagedorn 2001b, c, 2003a, b, c; Hagedorn and Gräfenhan 2002; Hagedorn and Weiss 2002; Hagedorn and Triebel 2003; Hagedorn et al. 2005, 2006) and the corresponding database applications are continuously optimized and extended.

DiversityDescriptions is used by several database projects on descriptive data (e.g., DEEMY, a project for the characterization and identification of ectomycorrhizae (www. deemy.de), GLOPP, a global information system for the biodiversity of plant pathogenic fungi, and LIAS, including LIAS light. The databases currently used for the subprojects LIAS names and LIAS checklists are based on parts of the information models of DiversityTaxonomy Names, SpecialIndexing and DiversityReferences. Currently, two freely available Web interfaces link the DiversityDescriptions databases of LIAS to the Internet for interactive identification: DeltaAccess Perl (a PERL script accessing DeltaAccess databases) and NaviKey 2.0 (a Java applet accessing DELTA flat files).

8.4 DATA STORAGE AND SERVICES

LIAS promotes the gathering, furnishing and administration of data by experts in a standard database system, which provides password-protected data and deposits for individual use (e.g., in context with ongoing monographic projects) that are made publicly accessible after the authors' assent. The core of LIAS is a list of more than 700 descriptors (characters, mostly multistate) that can be utilized in genus or species descriptions. The software architecture for the LIAS descriptive data is outlined by Figure 8.2.

Data entry and revision are performed online via database-generated HTML data entry forms (Figure 8.3). Considerable flexibility is built into data entry options in that modifiers and notes can be readily added. These data represent the source for database-generated natural language descriptions and online identification keys.

8.4.1 STRUCTURED DESCRIPTIVE DATA AND DEFINITIONS OF CHARACTERS

LIAS currently comprises more than 700 morphological, anatomical and biochemical data as well as distribution data. Characters (so-called descriptors) and character states, together with alternative wording (for natural descriptions) and definitions, are stored in the central

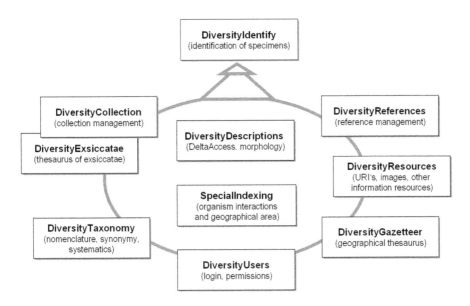

FIGURE 8.1 The Diversity Workbench database suite.

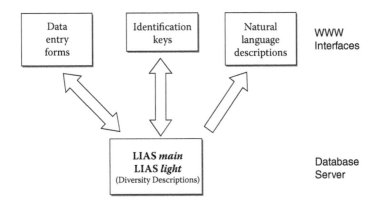

FIGURE 8.2 Software architecture for LIAS — descriptive data.

DiversityDescriptions database. Definitions are being elaborated as cooperation among the Botanische Staatssammlung München, the University of Bayreuth and Arizona State University as part of a project in the GBIF-D framework in close coordination with the project DEEMY (see earlier discussion; http://glossary.lias.net; http://deemy.de/). LIAS and DEEMY share structure and hierarchy of main descriptive characters and will use the same web interfaces and software tools for their online documentation.

8.4.2 LIAS Output

8.4.2.1 Natural Language Descriptions

Database-generated text with natural language descriptions (e.g., for usage in floristic or monograph projects) is provided in RTF, PDF or HTML formats (Figure 8.4).

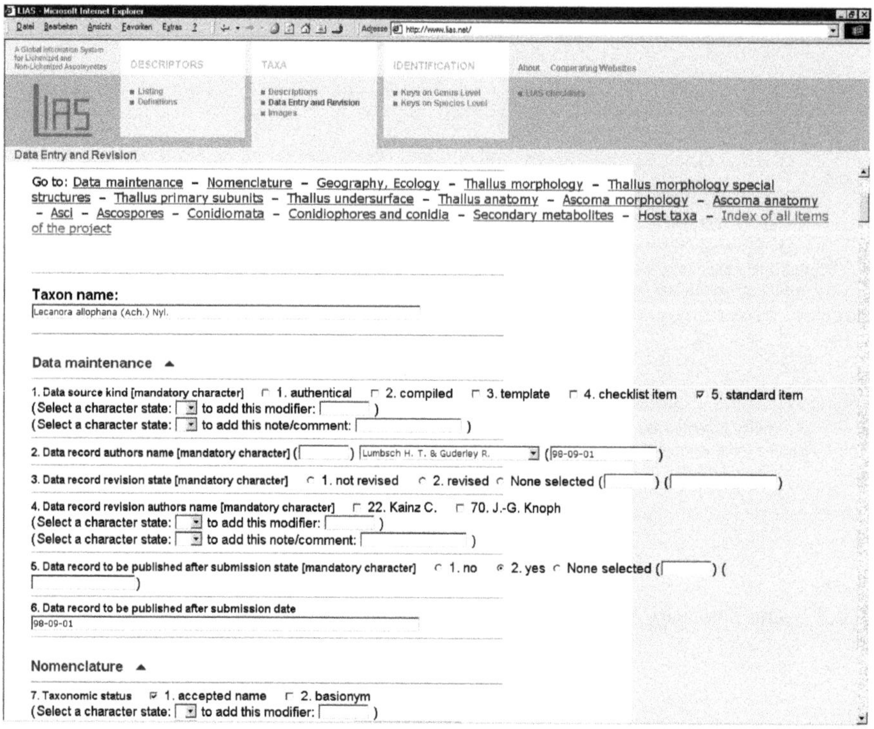

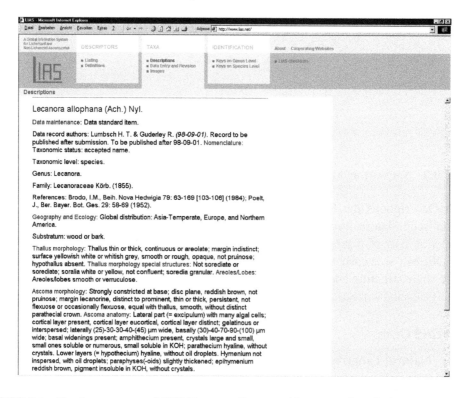

FIGURE 8.3 HTML form for data entry and revision.

FIGURE 8.4 Database-generated HTML output for natural language descriptions.

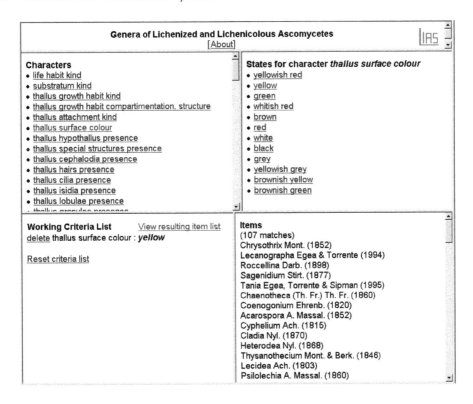

FIGURE 8.5   Online identification of lichenized and lichenicolous ascomycete genera.

8.4.2.2   Identification Keys

Interactive keys offer huge advantages over dichotomous keys in so far as one can utilize whatever characters are readily available and optimization strategies may be included so that keying within a set of, say, 500 species can be accomplished in relatively few steps. LIAS presents truly interactive Web-based keys for identification of ascomycetes (partly still at an experimental stage). Actually, a core key for all lichenized and lichenicolous genera (845) as well as various species level keys for 2000 species of 12 families of ascomycetes is available. Within the LIAS light project, an interactive key on 2600 lichens is currently being tested. This identification key utilizes only a small subset of 70 of the more than 700 characters of the LIAS descriptors list.

Currently, two free Web interfaces link the DiversityDescriptions databases of LIAS to the Internet for interactive identification: DeltaAccess Perl (Findling 1998) (Figure 8.5) and NaviKey, a Java applet for DELTA flat files (Bartley and Cross 1999). Both exist as beta versions and allow one to study the pros and cons of the different query modes. Meanwhile parts of the NaviKey code were reprogrammed and the functionality are being improved, e.g., by inclusion of additional query mode and the option of simultaneous character state selections (see http://www.navikeynet; Neubacher and Rambold 2005b onwards. Intkey is part of the CSIRO DELTA package and a stand-alone application that can be used as a so-called auxiliary application in the context of Web-based data provision (Dallwitz et al. 1995 onwards, 2000 onwards). With 2005 Intkey is abandoned in the framework of LIAS.

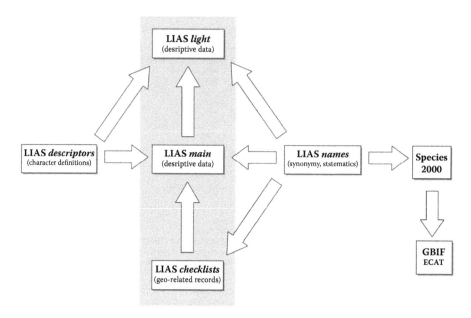

FIGURE 8.6   The modular structure of LIAS components.

8.5   LIAS SUBPROJECTS

In 1993, the LIAS project was already initiated with the intention of developing a multi-authored service for specialists for the entry and maintenance of descriptive genus- and species-level data to be used online and in the context of monographic works. Aside from this major goal, LIAS meanwhile gained importance with respect to (1) the general demand for rapid identification of organisms; (2) the demand for geospatial distribution of organisms; and (3) the demand for name pools. For enabling coverage of these aspects, three subprojects were set up in 2000 and 2001.

8.5.1   LIAS Light for Descriptive Key Data and Rapid Identification of Lichens

LIAS light (http://liaslight.lias.net) is embedded into the overall data structure of the core project and its data are stored in a corresponding way to the DiversityDescriptions database component (Figure 8.2 and Figure 8.6). The restriction of this submodule to a set of 70 characters allows more rapid data entry so that the majority of ascomycete species can be covered within the next few years. Data selection is optimized for the identification of lichenized groups. NaviKey and DAP are used as Web interfaces for the descriptive data of the LIAS core module (see previous discussion). By linking the data with information stored in LIAS checklists, it will be easy to integrate dynamically functions for country-specific preselection of taxa in the online identification keys.

8.5.2   LIAS Checklists for Spatial Data

The submodule LIAS checklists provides database access to spatial information on lichens and lichenicolous fungi for all 193 countries of the world and 300 additional geographical units at the subnational level (e.g., islands and states of larger countries). The geographic

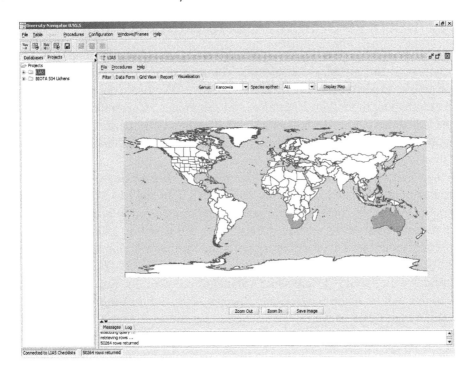

FIGURE 8.7   Status in 2004.

division follows in part the World Geographical Scheme for Recording Plant Distributions as proposed by TDWG and the *Getty Thesaurus of Geographic Names*. The checklist information is based on literature data and restricted to Europe, continental African countries, Southeast Asia, North America, Australasia, and Antarctica. It currently includes 14,135 taxon names respectively records (including synonyms). LIAS checklists subprojects shares layout standards and nomenclatorial compatibility with the other LIAS modules. Data are stored in the database component SpecialIndexing (Figure 8.1) and maintained using the database client Diversity Navigator (see following discussion). Visualization of the geospatial distribution of taxa is realized by a Web service via RPC SOAP, using the GIS system, GRASS, for the (status in 2004) generation of maps (Figure 8.7).

8.5.3   LIAS LICHEN NAMES FOR TAXONOMIC DATA

Actually, the database of LIAS names (using a SQL server version of DiversityTaxonNames) is storing taxon names, nomenclatural and taxonomic synonyms and concept names as used by LIAS subprojects. This information on names for lichens, lichenicolous fungi and powdery mildews is curated and expanded by experts. Names for other ascomycete taxa available from Index Fungorum as distributed in the Catalogue of Life (see e.g., Anonymous 2003 will be added). A web interface for query and browse LIAS names and classification is available under http://liasnames.lias.net. A web service is going to be established to provide lichen names to other web-based applications. This service will especially support the lichen projects within the German GBIF node for mycology (http://www.gbif-mycology.de) and facilitate access to LIAS content data in the context of the EU project Species 2000 Europa.

8.6   FURTHER DEVELOPMENTS AND REFERENCE TO GBIF

A major goal of LIAS is facilitating interaction and communication between experts of the lichenological and mycological scientific community and creating a network for data exchange among experts. More than 60 international scientists cooperated by compiling data and establishing more than 20 family-level subprojects. Currently, the system is going to be established as a data node associated to the German GBIF participant node for mycology (Rambold and Peršoh 2003; Triebel et al. 2003). LIAS is also included as global species database (GSD) in the frame of Species 2000.

Improved option concerning database architecture. For an improved performance on the Web, data of all LIAS modules, LIAS main, LIAS light and LIAS checklists, which are still stored in MS Access databases are going to be transferred to client-server databases, of which experimental versions already exist. They are based on identical data models as those of source databases DiversityDescriptions and SpecialIndexing. By installing a wrapper of the Species 2000 Wrapper Program that locates the Species 2000 relevant data and communicates with the SPICE hub using a standard protocol, the respective data from the LIAS modules will be made accessible to Species 2000 and thus to the Taxonomic Name Service (ECAT) of GBIF as well.

Taxonomic names. Taxonomic data are currently stored in all LIAS modules. In the near future, they will be stored and managed by the module LIAS names (see earlier discussion). By 2005 taxonomic data are managed and stored by the module LIAS names (see earlier discussion). For the classification part above genus level the co-operation with MYCONET (see http://www.fieldmuseum.org/myconet/outlie.asp) should be extended.

Descriptive data. As mentioned in GBIF Work Programme 2004 (Anonymous 2004), various biodiversity projects started attempting species-level synthesis of data from multiple sources and developing so-called species banks. LIAS is an example for such a species bank or global information system and has a strong focus on structured descriptive data assigned with definitions of morphological and other characters and their states. The exchange format currently used for data transfer from and into LIAS is DELTA. In the future, this might be replaced by the SDD format, which is currently developed as new interoperability standard for descriptive data (see TDWG working group: structure of descriptive data SDD http://wiki.tdwg.org/twiki/bin/view/SDD/WebHome).

The ensemble of descriptive data (including image data) and geospatial data, as stored in the LIAS system, represents material for a virtual mycota according to the concept referred to in the GBIF Work Programme (Anonymous 2004). It visualizes not only the geographic distribution of particular lichen or fungus species, but also understanding of the states of morphological and ecological characters within preselected taxa. Due to the principal option to assign the endangerment status of taxa to geospatial information, LIAS data are potentially applicative for the generation of lists and distribution maps of endangered species. With increasing quality of floristic data for the various regions of the world, the analysis of species richness at a regional and global scale appears a future option as well.

Collection data. The storage of historical specimen and record collection data as well as DNA sequence data is not the purpose of this databank. However, linking data of this type is possible by various technologies, as by direct interoperation between the database components DiversityDescriptions and DiversityCollection or by Web service functionality, using HTML forms or a database client as interface.

Improved options for data maintenance by remote access with database client. HTML-based Web interfaces for data entry and revision are suitable but not optimal, especially for data sets that include a large number of descriptors. Therefore, the client software Diversity Navigator (experimental version at http://www.diversitynavigator.net; Neubacher and Rambold 2005a onwards) is going to be adopted for optional direct database access. In addition to grid views for editing database contents, this platform-independent client (programmed in Java) also provides functions for querying data located in distributed database systems, as well as report functions for generating scripts for accessing SOAP RPC Web services, as shown in Figure 8.7.

ACKNOWLEDGEMENTS

We thank Gregor Hagedorn (Berlin), Markus Weiss (Munich) and Corinna Gries (Tempe) for valuable comments on the manuscript and Wiltrud Spiesberger (Munich) for preparing the figures. The activities of all LIAS data authors and revisers are gratefully acknowledged.

REFERENCES

Anonymous (2003) Catalogue of life. Indexing the world's known species. Species 2000 and ITIS Year 2003 annual checklist. CD-ROM, Species 2000.

Anonymous (2004) GBIF Work Programme 2004 (http://www.gbif.org/GBIF_org/wp/wp2004.

Bartley, M. and Cross, N. (1999) Navikey v. 2.0 (http://www.huh.harvard.edu/databases/legacy/navikey/index.html). New York. No longer accessible.

Dallwitz, M.J., Paine, T.A., and Zurcher, E.J. (1993 onwards) User's guide to the DELTA system: A general system for processing taxonomic descriptions, 4th ed. (http://biodiversity.uno.edu/delta/).

Dallwitz, M.J., Paine, T.A., and Zurcher, E.J. (1995 onwards) User's guide to Intkey: A program for interactive identification and information retrieval, 1st ed. (http://biodiversity.uno.edu/delta/).

Dallwitz, M.J., Paine, T.A., and Zurcher, E.J. (2000 onwards) Principles of interactive keys (http://biodiversity.uno.edu/delta/).

Eriksson, O.E., Baral, H-O., Currah, R.S., Hansen, K., Rambold, G., and Laessoe (eds) (2004). Outline of Ascomycota - 2004. Myconet 10:1-99.

Findling, A. (1998) DAP — Ein Web-Interface zu *DeltaAccess* (http://www.axel-findling.de/programs/dap/).

Hagedorn, G. (2001a) Making DELTA accessible: Databasing descriptive information. *Bocconea* 13: 261–280.

Hagedorn, G. (2001b) Documentation of the information model for DiversityDescriptions (1.8) (http://www.diversitycampus.net/Workbench/Descriptions/Model/2001-02-16/Diversity Descriptions_Model.html). Berlin.

Hagedorn, G. (2001c) Documentation of the information model for DiversityReferences (special indexing, 0.9) (http://www.diversitycampus.net/Workbench/References/Model/2001-05-08/DiversityReferencesData_Indexing_Model.html). Berlin.

Hagedorn, (G. 2003a) DiversityResources: Documentation of the information model (http://www.diversitycampus.net/Workbench/Resources/Model/2003-09-24/DiversityResourcesModel.html). Berlin.

Hagedorn, G. (2003b) DiversityReferences: Documentation of the information model (http://www.diversitycampus.net/Workbench/References/Model/2003-09-24/DiversityReferencesModel.html). Berlin.

Hagedorn, G. (2003c) DiversityGazetteer: Documentation of the information model (http://www.diversitycampus.net/Workbench/Gazetteer/Model/2003-09-24/DiversityGazetteerModel.html). Berlin.

Hagedorn, G. and Gräfenhan, T. (2002) DiversityTaxonomy (version 0.7 beta): Documentation of the information model (http://www.diversitycampus.net/Workbench/Taxonomy/Model/2002-03-05/DiversityTaxonomy_Model.html). Berlin.

Hagedorn, G. and Rambold, G. (2000) A method to establish and revise descriptive data sets over the Internet. *Taxon* 49: 517–528.

Hagedorn, G. and Triebel, D. (2003) DiversityExsiccatae: Documentation of the information model (http://www.diversitycampus.net/Workbench/Exsiccatae/Model/2003-09-24/Diversity ExsiccataeModel.html). Berlin.

Hagedorn, G. and Weiss, M. (2002) DiversityCollection information model (http://www.diversity campus.net/Workbench/Collection/Model/2002-11-15/DiversityCollectionModel.html). Berlin.

Hagedorn, G., Weiss, M., and Triebel, D. (2005). DiversityTaxonNames (version 1.0, 29 March 2005): documentation of the information model (http://www.diversitycampus.net/workbench/Taxonomy/Model/2005-03-29/DiversityTaxonNames.html). Berlin.

Hagedorn, G., Weiss, M., and Kohlbecker, A. DiversityReferences information model (version 2.0) (http://www.diversityworkbench.net/Portal/wiki/ReferencesModel_v2.0).

Nash, T.H. III, Gries, C., and Rambold, G. (2002a) Lichen floras: Past and future for North America. *The Bryologist* 105(4): 635–640.

Nash, T.H. III, Ryan, B.D., Gries, C., and Bungartz, F., eds. (2002b) *Lichen flora of the Greater Sonoran Desert region* 1. *Lichens unlimited*. Arizona State University, Tempe, 532 pp.

Neubacher, D. and Rambold, G. (2005a onwards) DiversityNavigator[R] — a Java rich client for accessing biodiversity databases (http://www.diversitynavigator.net).

Neubacher, D. and Rambold, G. (2005 onwards) NaviKey 4 — a Java applet and application for accessing descriptive data coded in DELTA (http://www.navikey.net).

Rambold, G. (1997) LIAS — The concept of an identification system for lichenized and lichenicolous ascomycetes. In *Progress and problems in lichenology in the nineties*, ed. R. Türk and R. Zorer IAL 3. *Biblioth. Lichenol.* 68: 67–72.

Rambold, G. (2001) Computer-aided identification systems for biology, with particular reference to lichens. In *Protocols in lichenology: Culturing, biochemistry, ecophysiology and use in biomonitoring*, ed. I. Kranner, R. Beckett, and A. Varma. Berlin, Heidelberg, 536–553.

Rambold, G. and Hagedorn, G. (1998) The distribution of selected diagnostic characters in the Lecanorales. *Lichenologist* 30(4–5): 473–487.

Rambold, G. and Peršoh, D. (2003) Structural optimization of the global information system LIAS by establishing a LIAS names server and expanding the Descriptors Workbench. In *Sustainable use and conservation of biological diversity. A challenge for society*. Symposium Report Part A, Berlin, 249–250.

Rambold, G. and Triebel, D. (1995 onwards) Genera of lichenized and lichenicolous Ascomycetes. LIAS. A global information system for lichenized and non-lichenized Ascomycetes http://www.lias.net.

Triebel, D., Braun, U., and Kainz, C. (2003) GLOPP — Erysiphales: Global information system for the biodiversity of powdery mildews. In *Sustainable use and conservation of biological diversity. A challenge for society*. Symposium report part A, Berlin, 214–215.

9 Linking Biodiversity Databases
Preparing Species Diversity Information Sources by Assembling, Merging and Linking Databases

Richard J. White

CONTENTS

ABSTRACT

This chapter discusses the ways in which information is being assembled in species diversity databases and combined to build comprehensive sources of biodiversity knowledge. Much of the chapter is drawn from the author's experiences with biodiversity data in various projects.

The manual processes by which data were merged in early projects are discussed, progressing through projects in which interoperability has been achieved by linking heterogeneous databases with very specific kinds of use in mind to the present possibilities for more general systems. Although techniques for interoperation are becoming increasingly sophisticated, issues of data quality and differences of expert opinion arise that can no longer be dealt with entirely manually, and ways to address some of these issues are also discussed.

9.1   INTRODUCTION

The process of systematic naming and recording of biological species has provided an evolving framework in which an increasingly diverse range of data types is organized. Biodiversity data consist of the name of a species and its synonyms, geographical and habitat data about where it is found, curatorial data about where reference specimens are stored, information about its appearance for identification purposes, its anatomy and chemical constituents, its genetic sequence, and so on.

As technology gradually advanced, biologists recorded these names and the associated data about each species or individual specimens in notebooks and on record cards and then assembled and published them in books and in databases. To pursue his or her research, a researcher investigating a particular group of organisms would have searched the literature, requested specimens from museums and herbaria or even visited them.

Because of the huge increase in the content and use of the World Wide Web, there is now great demand from users to access taxonomic and species diversity information on the Web and pressure on the taxonomic community to deliver it in a reliable and usable form. Inspection of individual specimens will always be necessary, especially by taxonomists, but as data pertaining to the specimens increasingly become available in electronic form, the need for such physical inspection by others will become less frequent.

In the following sections, I will describe progress in the construction, linking and sharing of biodiversity information in a series of species diversity database projects. First, I will use the ERMS project to illustrate some of the pitfalls encountered in the apparently simple task of assembling a Web site by importing data sets from a team of collaborators. Then, the process of assembling a species database by *merging* existing databases will be explored, using ILDIS as an example.

A large part of the chapter will then be concerned with issues that arise when *linking* online databases together through a gateway (Species 2000), including onward links to related information, checking the reliability of such links (LITCHI) and the prospects for 'intelligent linking' so that users are protected from some of the pitfalls associated with the nomenclature and subjective classification of species.

9.2   BUILDING SPECIES DATABASES

The increasing availability and accessibility of computers and general purpose database management software in the late 1970s and early 1980s led to the creation of a number of experimental database systems. One of these was the Vicieae Database (Adey et al. 1984), whose goal was to provide a fairly rich set of data on plant morphology and chemistry for every species in the Vicieae (vetches and peas), one of the tribes of the plant family Leguminosae (Fabaceae).

It was ambitious in the range of data types included, but was constrained by the limited structuring, editing and retrieval capabilities of the single-file mainframe data management system used at the time (Exir/Taxir). It was small enough that one person could take charge of data entry and editing. The database was used to generate a number of printed reports.

9.2.1 EUROPEAN REGISTER OF MARINE SPECIES

Later, the European Register of Marine Species (ERMS, http://erms.biol.soton.ac.uk) was an example of a project initially established primarily to construct a printed publication. It used a large team of participants including list editors who created approximately 100 separate species checklists for different taxonomic groups, mostly compiled as spreadsheets. These contained the scientific names and synonyms, higher classification and some optional fields, sometimes including information on geographical distribution, at least whether Atlantic or Mediterranean.

The format of the spreadsheets was insufficiently standardized. Required fields were sometimes separated and sometimes combined; for example, the fields genus, species, authority and date were often provided in a single field, necessitating an error-prone pattern-matching step to separate them. Geographical information was often presented in free text, again requiring particularly error-prone pattern matching.

Synonyms were provided in various ways: in separate records, linked by code or by name; in a separate field of the species record; mixed with other remarks, with various delimiters and separators; or even abbreviated or implied, as in 'Crella carnosa' (Topsent, 1904); (Lundbeck, 1910 as Grayella). The names of higher taxa were sometimes repeated in one or more fields of all the relevant species records or sometimes listed once only in a separate record, in various different formats, preceding the species that they included. This level of inventiveness among the list providers, coupled with a lack of documentation, suggested a belief that computers have great powers of intuition.

List importation and conversion was carried out in several stages: the Excel spreadsheets were exported to tab-delimited text files, which were interpreted by a Perl program with the aid of a manually prepared template for each list and imported into a client-server database (MySQL). The results of database queries were then passed through templates to generate RTF (rich text format, for the printed publication) (Costello et al. 2001) or HTML (for the checklist Web pages, see Figure 9.1). Although the data were held in a database, this was not initially seen as an end product, but it did support the Web-based taxonomic hierarchy (Figure 9.2). At the time of writing, conversion to a fully accessible database system was in progress.

9.3   ASSEMBLING DATABASES BY MERGING

Now that some species databases have been constructed, it becomes possible to assemble bigger ones by merging some existing databases together. A hypothetical diagrammatic example is shown in Figure 9.3.

9.3.1 INTERNATIONAL LEGUME DATABASE AND INFORMATION SERVICE

One of the earliest projects to create a computer database for a large group of organisms was ILDIS (http://www.ildis.org), the International Legume Database and Information Service. It was started in the 1980s when the wider availability of improved database management

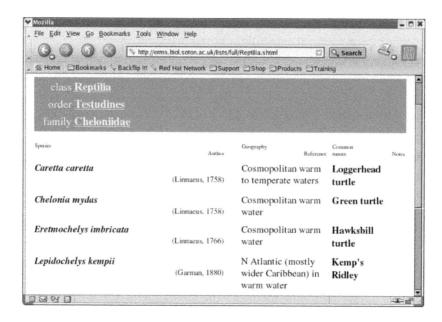

FIGURE 9.1 Part of the ERMS checklist page for Reptilia, showing an example of the static checklist format generated from the database in advance.

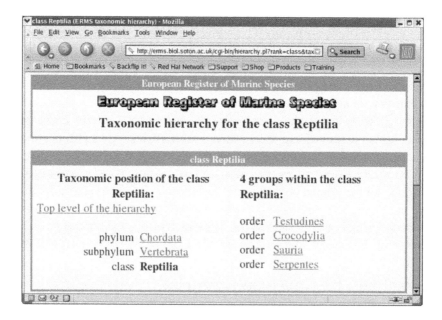

FIGURE 9.2 ERMS dynamic hierarchy page for Reptilia, generated from the database on demand.

systems such as dBase II on desktop computers made it possible to design and operate larger, more sophisticated data systems and to distribute extracts or copies of the entire database.

Unlike the Vicieae Database Project, ILDIS aimed to construct a database for the whole plant family Leguminosae. This encompasses about 20,000 species, perhaps 7–8% of the world's flowering plants, including peas, beans and related pod-bearing plants, many of

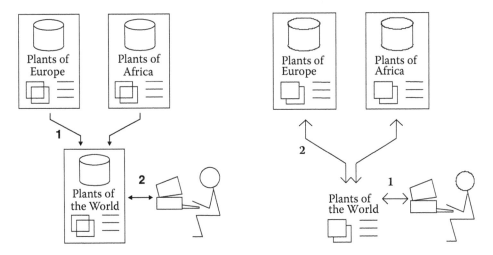

FIGURE 9.3 Merging databases. 1. The original databases are physically copied into a new combined database. 2. The user interacts with the new combined database.

which are of economic value. Its goals include building, maintaining and enhancing the ILDIS World Database of Legumes (Bisby et al. 2004) and designing and providing services from it to users, including an online Web database (LegumeWeb) and through the Species 2000 gateway described later.

The project began, like many others, as an informal cooperation among a number of specialists, in this case taxonomists and applied biologists with a scientific interest in the family. It is now an international collaborative project with a coordinating centre; approximately 10 regional coordinators responsible, typically, for a whole continent (or large area such as the former USSR), who had already begun, in many cases, to assemble regional data sets; and 30 taxonomic coordinators responsible for particular taxonomic or regional sectors of the family.

Although a purpose-built software package (Alice) for building species databases was used (see White et al. 1993), hindsight has shown that this was lacking in crucial features now considered essential, such as the capability for distributed querying and editing and for merging separately constructed databases. Nevertheless, the merging of separate regional checklists was central to the design of the ILDIS data management process. The lack of software support for this was alleviated by a reduced range of data types, compared to the Vicieae database, together with a data entry regime that took place at one site, with data editing and checking performed on paper print-outs at regional centres.

The core taxonomic checklist is nearing completion and provides a consensus taxonomy — a unified taxonomic treatment or backbone to which various kinds of additional data may be attached. Version 7.03 of the ILDIS World Database of Legumes comprised 19,554 taxa, of which 15,574 are species, 1587 subspecies and 2393 varieties. In addition to the 19,554 accepted names, there are 20,101 synonyms and various misnomers, making 39,655 names altogether.

9.3.2   ILDIS LᴇɢᴜᴍᴇWᴇʙ

Because the Alice software that supports the ILDIS database is a single-user DOS program and therefore provides its user with only a local user interface, it is unsuitable for hosting a

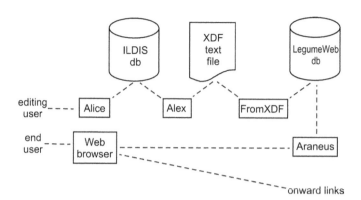

FIGURE 9.4 Conversion of the ILDIS database to LegumeWeb. The upper user is the ILDIS database editor, who also initiates the conversion process, shown with solid lines. The lower user is an end-user of the LegumeWeb database, whose Web interactions are shown with broken lines.

Web database. It was necessary to provide a separate online version of the database, called LegumeWeb (http://www.ildis.org/legumeweb/). The Web interface allows the database to be enriched, partly by means of images and partly through hyperlinks to other data sources as described next.

Periodically, LegumeWeb is refreshed by performing a series of conversion steps. First, Alex (a program in the Alice suite) is used to export the ILDIS database to a text file, then the specially written program FromXDF is used to convert this to a MySQL database, from which a Web user interface can be run using another program, Araneus (Figure 9.4). The LegumeWeb Web-based user interface demonstrates a couple of important features: two-stage access with synonymic indexing and its use as a gateway to external information by means of onward links (direct species name links) to further data sources.

In the first step, the user types in a name (Figure 9.5), which may be incomplete. LegumeWeb responds by showing a list of the species names that match the user's specification (Figure 9.6). In the second step, the user chooses one of the species names provided, which may be a synonym or an accepted name. In this example, the user chooses *Abrus cyaneus*, a synonym for *A. precatorius*, and LegumeWeb responds by showing a standard set of information about the chosen species *A. precatorius* (Figure 9.7).

9.3.3 SYNONYMIC INDEXING

Automated synonymic indexing is the process ensuring that, when a user attempts to search for information using a synonym or a name that has often been used erroneously in the past, the user is informed about this and directed to the correct information listed under the currently accepted name of the species. Essentially, this is the translation of a name into the corresponding taxon and cannot always be done automatically. In some cases at least, it is desirable to offer an explanation to the user. Conversely, it may also be informative, when the taxon is displayed, to show the synonyms under which it has often been listed in older works.

There are several types of synonyms and names used in error. For the purpose of the synonymic indexing process, they can be broadly classified into two classes: those that can unambiguously be interpreted as referring to a particular taxon and those ambiguous

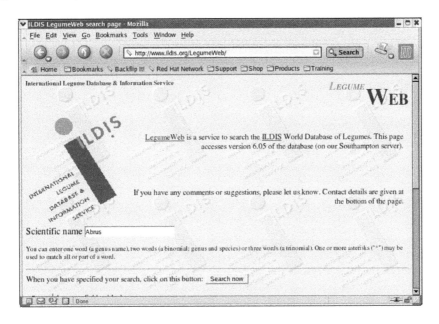

FIGURE 9.5   ILDIS LegumeWeb search form. The user has entered the search term 'Abrus'.

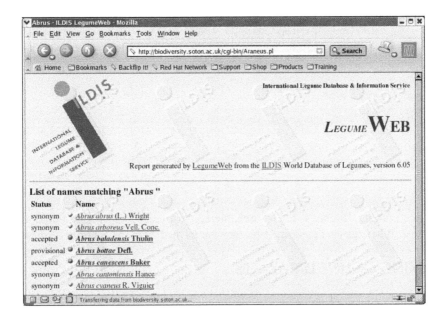

FIGURE 9.6   ILDIS LegumeWeb returns the names that match the string 'Abrus'.

names that cannot. The latter are of three main kinds: *pro parte* synonyms, which refer to taxa that have now been split up, so the name could refer to any or all of these; homonyms, which by definition refer to more than one species, although they may be distinguishable if the authority name is quoted; and misapplied names, in which case it may not be known whether the name is being used in its correct or incorrect sense.

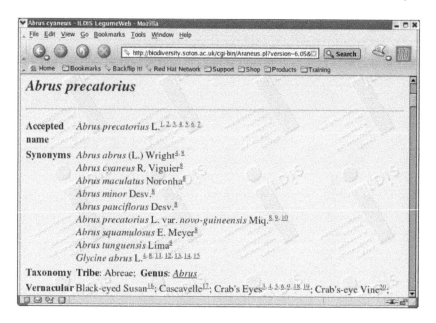

FIGURE 9.7   Top of the ILDIS LegumeWeb data display for the species *Abrus precatorius*.

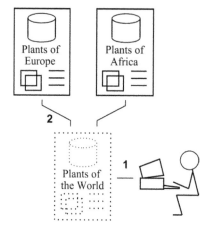

FIGURE 9.8   Linking databases. 1. The user interacts with an access system that does not itself contain data. 2. When the user requests data, they are fetched from the appropriate database.

9.4   LINKING ONLINE DATABASES

Instead of physically merging them, as previously described, databases that can be accessed through an online point of entry could be linked to a portal (initially just a Web site), which could then act as a user interface to provide a single entry point into a series of such databases. These could be seen by the user as a larger virtual database (Figure 9.8).

9.4.1   SPECIES 2000

The existence of species databases such as ILDIS and FishBase (http://www.fishbase.org), whose data are broadly comparable, led to the idea of linking them to construct a virtual checklist or catalogue of all species of organisms. The Species 2000 project (http://www.

sp2000.org) is an international collaborative project to provide access to such an authoritative and up to date virtual checklist. Its system architecture comprises a distributed array of global species databases (GSDs) arranged 'side by side' so that all groups of organisms are included and accessed at present through a Web interface and a Web Service.

The standard data provided by the GSDs comprises the information about a species that Species 2000 wishes to deliver: its accepted name; frequently occurring synonyms and common names, all with references; its family or other appropriate recognized higher taxon; a comment field that can be used for different purposes in different GSDs; a field to indicate when and by whom the species entry was last checked or scrutinized; optional geographical distribution information; and one or more URLs linking to further data sources for this species, using onward links as described later.

A complete array of such GSDs will take some time to be completed, but the principle was demonstrated successfully in a simple early proof of concept prototype, still available at the Species 2000 Web site. The general architecture has several advantages: the single portal removes the need for a user to discover and address every GSD separately; a single operation can search all the databases simultaneously if desired; and data can be standardized to facilitate retrieval, comparison and collation.

Such a catalogue could be explored in two main ways: by browsing through the species names, which is facilitated by their classification into a taxonomic hierarchy of nested larger groups, or by searching for names matching a search pattern. Having located the name of a species, the user would typically want to view more information about that species. In the first prototype, the latter type of interface was implemented, in which the user, who is assumed to have a complete or partial scientific name in mind for the search, interacts with the Species 2000 Web portal in several stages, similar to those used in LegumeWeb:

- The user enters a search string, which represents the species he or she has in mind more or less completely and chooses which GSDs are to be searched.
- The GSDs searched return lists of names that match the user's search string and from which the user, perhaps on the basis of additional knowledge, chooses the name of interest.
- The GSD that contained the selected name then provides a page of basic information about the species whose name was selected.
- Optionally, this page may contain hyperlinks retrieved with the other data from the GSD in question that point to further information about the species elsewhere on the Web; the user can follow these links at will. Thus, the databases linked to the gateway can in fact be arranged in tiers, with a primary array of GSDs and a secondary array of sources of further information.

This approach has become known as the 'Catalogue of Life', especially since Species 2000 began cooperating with the North American governmental organization ITIS (http://www.itis.usda.gov), which has a related goal of providing a catalogue of species names.

9.4.2 THE SPICE PROJECT

The original Species 2000 implementation of its gateway consisted of a single, simple CGI program invoked when the user completed a Web form (http://biodiversity.soton.ac.uk/

sp2000/CAS2.html) representing the first stage described previously. The program consulted a list of the GSDs available (only a handful at that time) and returned to the user's browser a Web page containing frames, one frame per GSD. Each frame contained a URL, which called a CGI program or 'wrapper' to search one of the GSD databases and return a list of names matching the search string as a Web page to be displayed in that frame.

The use of frames seriously compromised scalability, since the number of frames that can be presented simultaneously in a Web page is severely limited by practical constraints. Also, the nature of frames meant that no further control, monitoring or interaction with users' activities was possible: they were left on their own. In addition, the implementation was limited in other ways: metadata about the GSDs available and their properties was hard coded into the CGI programs, and the delivery of the data in HTML did not encourage the use of automatic software clients for further uses of the data.

The SPICE Project (http://www.systematics.reading.ac.uk/spice/) included researchers at the Universities of Cardiff, Reading and Southampton, working with the Royal Botanic Gardens, Kew, the Natural History Museum, London, Species 2000, NIES (Japan), BIOSIS UK, and ILDIS. Its goal was to remove many of these obstacles so that the system could potentially scale up to hundreds of GSDs and thousands of connected users searching for names, as well as be able to overcome heterogeneity in the database systems while retaining their autonomy (Jones et al. 2000b; Xu et al. 2001, 2002).

This was achieved by the implementation of an integrated middleware layer or hub (the Common Access System or CAS) to replace the CGI programs. It acts as a mediator, passing user requests to the appropriate GSD or GSDs and assembling responses from the GSDs before presenting the results to the user. Interoperability was achieved by using wrappers to map the GSDs into a common data model. In order to locate GSDs, a limited amount of metadata that can easily be edited as new GSDs join the federation is held by the CAS. Also, the mediator role of the CAS allows it to create and use indexes and caches that allow the appropriate GSDs to be selected and the names to be found more efficiently.

9.4.3 THE COMMON DATA MODEL

Different people are building the various components of the system: the GSDs, the wrappers, the CAS and the user interface. We need to ensure they all have a common understanding of the data to avoid errors in data transfer.

We use a common data model (CDM), which is a document to define the information being passed to and fro. It is human readable, not machine readable, but is a reference source used to create specific machine-readable implementations for Corba (IDL), CGI/XML (DTD, XML Schema), etc. It defines the input (request) and output (response) for six fundamental operations which the Species 2000 and SPICE systems need to be able to carry out in order to achieve their design goals. The operations are referred to as request types 0 to 6:

Type 0: get the CDM version that the GSD wrapper complies with.
Type 3: get information about the GSD.
Type 1: search for a name in a GSD (stage 1 of the user's interaction when searching for a species by name).
Type 2: fetch the standard data about a chosen species (stage 2 of the user's interaction when searching for a species by name).

Type 4: move up the taxonomic hierarchy towards the 'root' of the tree, which is actually the highest level taxon (used when a user is browsing the species list).

Type 5: move down the taxonomic hierarchy towards the species level (used when a user is browsing the species list).

9.5   LINKS TO EXTERNAL SPECIES DATA

A wide range of biologists, environmental scientists and other professionals now rely on access to biodiversity information in order to carry out analyses, construct models and write reports. A variety of biological databases and other data sources that provide data useful for studies in biodiversity are currently available and under construction. As described earlier, a more complete catalogue (or checklist) of the names of all species of organisms can be assembled by linking together separately constructed databases containing the *same* kind of information about *different* groups of organisms, such as bacteria, fish, groups of plants, and so on.

In contrast to such links within a managed project, species databases and GSDs, such as ILDIS, and federated gateways to these systems, such as Species 2000, also envisage providing so-called onward links from their data to external sources of related data. In this way they can serve not only as catalogues that only list species but also as indexes that potentially provide direct access to all species information on the Internet. Species 2000 has long had plans to provide links to take a user from one of its species entries (provided by a GSD) to further sources of information about that particular species.

In this way, databases containing *different* kinds of information about the *same* group of organisms are linked together. A taxonomist writing a monograph, a conservationist assessing a threatened species, or a plant breeder would be able to use different kinds of information about the species from different sources.

In the case of ILDIS, for example, additional data items attached to the core checklist may be managed by ILDIS, such as maps, images, annotations, generic summaries, higher taxa and the Phytochemical Database, used to create the Phytochemical Dictionary of the Leguminosae (Southon et al. 1994), or data may be linked to by ILDIS but managed by others, as described later.

9.5.1   ONWARD LINKS

Originally, ILDIS relied on a DOS-based program (Alice) that had no easy way to handle such links, but the Web now provides ways to do this. The user may follow a hyperlink from the LegumeWeb page (Figure 9.9) to some other data source for further information. For example, a user of LegumeWeb may choose to go to W³Tropicos at Missouri Botanical Garden to see more information about the same species (Figure 9.10), and maybe proceed to obtain further information (Figure 9.11). In this way LegumeWeb acts as a gateway to other information about legume species.

9.5.2   CHECKING THE RELIABILITY OF LINKS

As biodiversity data become increasingly available, data quality problems become more noticeable. To some extent, this is a good thing: data quality problems are now more likely

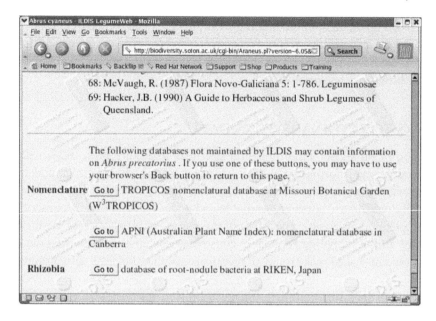

FIGURE 9.9   Bottom of the ILDIS LegumeWeb data display for the species *Abrus precatorius*, showing onward link buttons to visit external data sources.

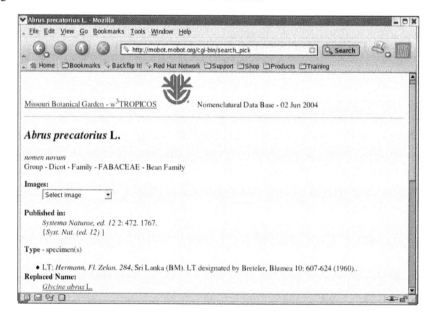

FIGURE 9.10   The destination of an onward link: W³Tropicos at Missouri Botanical Garden.

to be detected than in the past because conflicts between data sets may be found when they are used in combination that would not otherwise have been detected. But whereas there are well-established techniques for cross-platform interoperation, techniques for semantic interoperation, especially for situations where data quality is less than ideal, are still in need of continued research. Progress is required in areas such as these if we are to realize the full potential of biodiversity resource networks to address research issues such as the effects of global climate change.

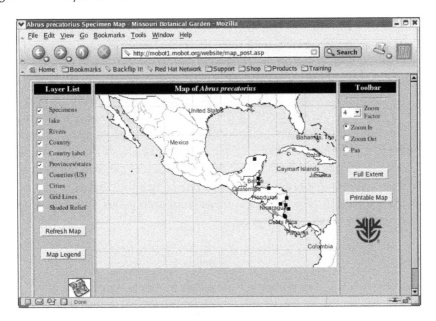

FIGURE 9.11 The distribution of *Abrus precatorius* specimens reported by W³Tropicos.

In particular, an issue that must now be faced is the question of the reliability of the very names that we are using as keys and links in building and connecting databases. Although organisms were observed and studied before his time, Linnaeus established a procedure for the systematic cataloguing of animals and plants. He standardized the system of giving species two-word Latinized scientific names. Later workers developed rules for handling synonyms and name changes because they were faced with the issues of duplicated names and changing opinion as to the classification of the species and their circumscription — for example, how broad a group of organisms the name should apply to and hence whether species should be split or merged.

A key principle, which has a number of consequences, is that in many cases a species retains its original name even when its circumscription has changed. This reduces the need for new names to be created, but introduces ambiguity as to exactly what concept of the species the name refers to. This is important because the scientific name of a species is the key to all the information known about it.

Nomenclatural conflicts are certain kinds of errors or discrepancies in the presentation of a checklist that can be detected automatically by the application of rules and then, if desired, acted upon to try to correct them automatically or by the intervention of a taxonomist or skilled editor.

In the following example of extracts from two databases or checklists, the species listed in the two extracts appears to be the same, but there is a difference of opinion about which name should be used for it. The synonyms are unambiguous as to the species to which they refer.

Database A	Database B
Caragana arborescens **Lam.** (accepted name)	*Caragana sibirica* **Medikus** (synonym)
Caragana sibirica Medikus (accepted name)	*Caragana arborescens* Lam. (synonym)

Taxonomic uncertainties and overlaps can be inferred from some of the discrepancies detected by the rules just described, with some additional rules. For example, two checklists or databases can each be internally correct and consistent, but when considered together they may reveal that a species in one database is treated differently in the other. For example, it may have been split, combined with another, or overlap in a more complex manner.

In the next example, the existence of the identically named species *C. crista* in both databases is misleading; rather than being identical, closer inspection of *C. bonduc* in database B reveals that *C. crista* has been split into two species, so the one whose accepted name is *C. crista* may not be identical in circumscription to the one labelled *C. crista* in database A. The 'p.p.' is an abbreviation for *pro parte*, meaning that only part of the original *C. crista* is referred to. The name *C. crista* is an ambiguous synonym in database B because it may refer to either of two species.

Database A	Database B
Caesalpinia crista **L.** (accepted name)	*Caesalpinia crista* **L.** (accepted name)
	Caesalpinia bonduc **(L.) Roxb.** (accepted name)
	Caesalpinia crista L., p.p. (synonym)

9.5.3   THE LITCHI PROJECT

Whether in merging data sets to construct a species database like ILDIS or in linking from one data set to another, it is necessary to ensure that the species concepts in the different databases do not conflict and to take the appropriate action if they do. The Litchi Project (http://litchi.biol.soton.ac.uk) was a collaboration among the Universities of Cardiff, Reading and Southampton to design and test a rule-based tool for the detection and repair of conflicting nomenclature and taxonomic circumscription in cases such as these, which can arise when merging or linking data in taxonomic databases.

We modelled the knowledge integrity rules in a checklist or taxonomic treatment. The knowledge is implicit in the assemblage of scientific names and synonyms used to represent each taxon. Although we often refer to checklists, which could be separate documents, these are equivalent to the set of names and synonyms held in a database, which may also hold other information about the taxa contained.

A prototype demonstration tool for merging checklists and testing their integrity was implemented as 'Litchi version 1' (Jones et al. 2000a). It applied a series of rules written in Prolog to a checklist, or to two checklists, to detect and report errors in the construction of the checklists — for example, a name that occurred twice in circumstances where this would be an error, for a variety of reported reasons. The system relied on several commercial software packages and was therefore cumbersome and difficult to port, but successfully demonstrated some of the principles and could be employed as a data quality checker. Practical uses include helping a taxonomist to detect and resolve taxonomic conflicts when merging or linking two databases.

9.5.4   'INTELLIGENT' SPECIES LINKS

Given that Litchi showed that it is possible to detect many cases of potential taxonomic conflict when linking species databases, how can such links be managed? There are a number of choices in the ways links may be made and handled. A hyperlink may be:

- a *general* link to a database or project's home page (not very useful because the user has to search for the species again);
- a *fixed* link to a known static species page;
- an *unchecked* name link to search a database for the required name (as in the present LegumeWeb prototype), which may be absent or conflict in scope;
- a *checked name* link to a known name entry in a database;
- a *checked species* link to a known species database entry; or
- an *'intelligent'* or *concept* link, with some metadata to allow an intelligent link robot to check or follow the link and take appropriate action.

Links from GSDs to other species information sources (SISs) can be established by discovery (performed by the GSD organization) or by registration (performed by the SIS organization and recorded by the GSD). Such links and any associated metadata need to be stored and must be checked regularly, possibly with automatic assistance, to ensure that they remain working and appropriate. Any checking required to establish the validity of a link may be done in advance (batch checking), when the user displays a source page (containing the link) or when the user follows the link to the target page.

9.5.5 CROSS-MAPPING

A system for managing and assisting users to navigate species links intelligently would maximize the potential of the plethora of species-based catalogues, indexes and rich species resources currently being assembled all over the world. So how can intelligent links be made to work, especially in the difficult cases where a species in one database does not have an exact match in the other? One way is to create and maintain cross-maps, which are more general than single species links in that they can describe how one or more taxa in one resource (such as the Species 2000 index) relate to one or more taxon concepts in another, perhaps extending the basic concept relationships listed by Geoffroy and Berendsohn (2003).

Cross-maps may be created and maintained in various ways: manually by experts; by monitoring the behaviour of users following species links; or automatically or semi-automatically by an enhanced Litchi tool (discussed in the following section), possibly assisted by analysing data sets describing the taxa, when sufficient such data are available, using the usual species taxonomy tools (phenetic and cladistic analyses). Future development of Litchi may allow the user to browse or analyse attached data to be organized, merged and used to support conflict resolution in this way.

Cross-maps may be held by individual GSDs, describing how to link their species to selected related resources that may use different taxonomy, made available in a cross-map server accessible to multiple projects, or stored in a repository for use by an intelligent linking engine, as in the extended SPICE hub to be built for the Species 2000 Europa project.

9.6 SPECIES 2000 EUROPA

Species 2000 Europa (http://sp2000europa.org) is an EU-funded project to support the development of the central hub of Species 2000 and surrounding databases in a European context. Specifically, it includes the development of an improved version of the SPICE hub software, support for linking to more databases, and an explicit mechanism for dealing with the differences in taxonomy and nomenclature between regional and global databases. Notionally, there will be separate regional and global hubs to which several regional and many global databases will be respectively attached. The user will deal with one or other of these hubs, thus receiving a regional or global view of the taxonomy and nomenclature. The user will be able to step across from one view to the other.

In order to build the cross-maps necessary to support this dynamic linking and user navigation, it is necessary to detect and record the various kinds of species concept relationships that exist between the databases. Therefore, a new version of the Litchi software, Litchi version 2, is being designed and built as part of the Species 2000 Europa project. The new version is intended to be much more portable and is written mostly in Java, with a redesigned database accessible using SQL. It has two levels of rules. The first level, roughly comparable to the Litchi 1 rules, detects data quality issues and establishes the correspondences between names in the checklists. The second level rules interpret these name associations in terms of species concept relationships and construct a cross-map that documents these relationships in an operational form so that checklist portals and browsing software can assist a user who wants to move from a species in one source to the corresponding species in another or view their data together.

9.7   CONCLUSIONS

In the foregoing sections I have described systems that were built originally to meet quite specific needs — ranging from a simple architecture for information retrieval from a single, hand-crafted database to the provision of access to a common set of data distributed across a federation of complementary databases — and their use as an index to further information sources. I have mostly used projects in which I have been directly involved, in order to illustrate the progress that has been made in the increasing sophistication of species diversity data systems and their interoperation.

In such systems the type of information provided was predetermined and the system was designed to provide a single point of access for a limited range of users and uses. But, ideally, scientists should have an environment — a problem-solving environment (PSE) — in which all the data of interest to them are accessible and within which analyses can be performed using a variety of tools provided within the environment; this should complement the scientists' working practices. Moreover, the PSE should be extensible as new resources become available and provide mechanisms for the discovery of such resources.

The BiodiversityWorld project (Jones et al. 2003; White et al. 2003) represents a step in this direction and is described in more detail in Chapter 7. It is intended to be of general interest to the biodiversity informatics community, a community that has accumulated many specialized collections of data and tools, but could benefit from sharing these resources more efficiently. It is an e-science pilot project that aims to achieve interoperability among a diverse range of databases and software by allowing users to link these together in workflows. Crucial in such linking will be techniques to handle the ambiguity of names as described previously.

Such developments might be on the Web, as with the current Species 2000 systems, or on the forthcoming Grid. Andrew Jones describes the potential of the Grid for biodiversity use in Chapter 7.

ACKNOWLEDGMENTS

I am delighted to have this opportunity to thank Professor Frank Bisby, School of Plant Sciences, The University of Reading, and Professor Alex Gray and Dr Andrew Jones, School of Computer Science, Cardiff University, for the very significant contributions they have made to the ideas and projects described in this chapter. The Litchi and SPICE for Species 2000 projects were funded by grants from the UK EPSRC/BBSRC Bioinformatics Special Initiative. ERMS was initially funded by the MAST programme of the European Union, and Species 2000 Europa was funded by a grant held under EU Framework V. BiodiversityWorld was funded by a BBSRC grant. I should also like to thank the Systematics Association and the Linnean Society for arranging and supporting the conference and this resulting volume.

REFERENCES

Adey, M.E., Allkin, R., Bisby, F.A., White, R.J., and Macfarlane, T.D. (1984) The Vicieae database: An experimental taxonomc monograph. *Databases in Systematics*, ed. R. Allkin and F.A. Bisby. Systematics Association, special volume 26, 175–188.

Bisby, F.A., Zarucchi, J.L., Schrire, B.D., Roskov, Y.R., and White, R.J., eds. (2004) *ILDIS world database of legumes*, ed. 8. ILDIS, Reading. (Online database and electronic publication [LegumeWeb service]: http://www.ildis.org).

Costello, M., Emblow, C., and White, R.J. (2001) *European register of marine species*. Muséum National d'Histoire Naturelle, Paris, 463 pp.

Geoffroy, M. and Berendsohn, W.G. (2003) The concept problem in taxonomy: Importance, concepts, approaches. In *MoReTax: Handling factual information linked to taxonomic concepts in biology*, ed. W.G. Berendsohn. Schriftenreihe für Vegetationskunde, volume 39, Federal Agency for Nature Conservation, Bonn, Germany, 5–14.

Jones, A.C., Sutherland, I., Embury, S.M., Gray, W.A., White, R.J., Robinson, J.S., Bisby, F.A., and Brandt, S.M. (2000a) Techniques for effective integration, maintenance and evolution of species databases. In *12th International conference on scientific and statistical databases*, ed. O. Günther and H.-J. Lenz. IEEE Computer Society Press, 3–13.

Jones, A.C., Xu, X., Pittas, N., Gray, W.A., Fiddian, N.J., White, R.J., Robinson, J.S., Bisby, F.A., and Brandt, S.M. (2000b) SPICE: A flexible architecture for integrating autonomous databases to comprise a distributed catalogue of life. In *11th International conference on database and expert systems applications* (LNCS 1873), ed. M. Ibrahim, J. Küng, and N. Revell. Springer Verlag, Berlin, 981–992.

Jones, A.C., White, R.J., Pittas, N., Gray, W.A., Sutton, T., Xu, X., Bromley, O., Caithness, N., Bisby, F.A., Fiddian, N.J., Scoble, M., Culham, A., and Williams, P. (2003) BiodiversityWorld: An architecture for an extensible virtual laboratory for analysing biodiversity patterns. *In Proceedings of the UK e-science all hands meeting*, Nottingham, UK, EPSRC, 759–765.

Southon, I.W., Bisby, F.A., Buckingham, J., Harborne, J.B., Zarucchi, J.L., Polhill, R.M., Adams, B.R., Lock, J.M., White, R.J., Bowes, I., Hollis, S., and Heald, J. (1994) *Phytochemical dictionary of the Leguminosae*. Chapman & Hall, London, 1580 pp.

White, R.J., Allkin, R., and Winfield, P.J. (1993) Systematics databases: The BAOBAB design and the ALICE system. In *Advances in computer methods for systematic biology: Artificial intelligence, databases, computer vision*, ed. R. Fortuner. The Johns Hopkins University Press, Baltimore, MD, 297–311.

White, R.J., Bisby, F.A., Caithness, N., Sutton, T., Brewer, P., Williams, P., Culham, A., Scoble, M., Jones, A.C., Gray, W.A., Fiddian, N.J., Pittas, N., Xu, X., Bromley, O., and Valdez, P. (2003) The BiodiversityWorld environment as an extensible virtual laboratory for analysing biodiversity patterns. In *Proceedings of the UK e-science all hands meeting*, Nottingham, UK, EPSRC, 341–344.

Xu, X., Jones, A.C., Pittas, N., Gray, W.A., Fiddian, N.J., White, R.J., Robinson, J.S., Bisby, F.A., and Brandt, S.M. (2001) Experiences with a hybrid implementation of a globally distributed federated database system. In *2nd International conference on Web-age information management* (LNCS 2118), ed. X.S. Wang, G. Yu, and H. Lu. Springer Verlag, Berlin, 212–222.

Xu, X., Jones, A.C., Gray, W.A., Fiddian, N.J., White, R.J., and Bisby, F.A. (2002) Design and performance evaluation of a Web-based multitier federated system for a catalogue of life. In *Proceedings of the fourth international workshop on Web information and data management* (WIDM 2002). ACM, 104–107.

10 Priority Areas for Rattan Conservation on Borneo

Jacob Andersen Sterling, Ole Seberg, Chris J. Humphries,
F. Borchsenius and J. Dransfield

CONTENTS

ABSTRACT

In recent decades conservation of biodiversity has been high on the international agenda, following the establishment of the Convention on Biological Diversity in 1992 (United Nations 1992a, b). International agreements have called for conservation efforts that prioritize areas particularly rich in biodiversity or in other ways unique. Parallel to the political debate, a scientific debate on how to prioritize conservation efforts is ongoing. This chapter documents the diversity and distribution patterns of the rattans on Borneo, assesses their conservation status and identifies priority areas for rattan conservation. It is the first study of any plant group on Borneo using a specialized analytical database, WORLDMAP. A total of 5045 rattan records were gathered from eight different herbaria and through fieldwork. Two thirds of the 144 rattan taxa on Borneo are endemic. At least 18.8% of the rattans on Borneo and at least 23.5% of the endemics are threatened. Three rattans are critically endangered. Patterns in taxon richness indicate that the northern and north-western areas harbour more taxa than those in the south and the east, and patterns in range-size rarity indicate that endemism hotspots are found in the south-western Sarawak, Brunei and the central parts of Sabah. The protection of rattans provided by the current set of reserves is no better than a random selection of areas. It is shown that a complementarity method is the most efficient in identifying priority areas for rattan conservation on Borneo. Using complementarity, 23 of the 1087 grid cells in WORLDMAP are required to represent all taxa at least once and, where possible, in a grid cell with more than 10% natural vegetation, 26 grid cells are required.

10.1 INTRODUCTION: HOW TO IDENTIFY PRIORITY AREAS

Perhaps the first step in the determination of reserve areas is to understand the goals of the exercise, to understand the methods and to create the database accordingly. Areas designated as reserves are often not selected to maximize biodiversity conservation (Pressey 1994) and thus they may be inefficient in doing so. To fulfill international obligations regarding *in situ* conservation of biodiversity and, in particular, to address recommendations to support initiatives aimed at areas particularly important for biodiversity, it is crucial that conservation planners apply methods of area selection developed to ensure efficient protection of biodiversity.

The term 'reserve' is here used in a broad sense to describe areas under a wide range of *in situ* biodiversity protection measures, as suggested by Pressey et al. (1993). Design and management of reserves and reserve networks are extremely important in the debate on how to ensure long-term biodiversity conservation. This section will, however, solely address the issue of geographical location of reserves. It will focus the analytical database on 'where first' rather than 'how' in conservation (Pressey et al. 1993).

10.1.1 THE CURRENCY OF BIODIVERSITY AND ITS SURROGATES

'Biodiversity can be seen as the irreducible complexity of life, and thus no objective measure or valuation of biodiversity is possible' (Williams 1998). However, in setting priorities for reserve selection, it is fundamental to be able to state the value (the goal) of different areas in order to be able to compare them and to prioritize them. It is instrumental that there is agreement on a fundamental biodiversity 'currency'. As mentioned earlier, a commonly used unit to describe or measure biodiversity is the species. However, the definition of biodiversity used in the CBD refers to the variety of all life at all levels, including between and within species (Glowka et al. 1994). This definition implies that the fundamental currency of biodiversity may not be found at the species level, but rather at the genetic level as the expressible characters of organisms and of the systems in which they play part (Williams 1998).

Defining characters as the fundamental currency for biodiversity value and giving the characters equal weights means that the units in which they are measured (individuals, species, ecosystems or areas) naturally may have different values because they contribute different numbers or combinations of characters (Humphries et al. 1995). The task of conservation planners should be to maximize character value within the networks of nature reserves (Williams 1998).

This is a rather theoretical approach that, to be used directly, implies that all characters of all organisms should be counted. In practical conservation planning such data will hardly ever be available. It is therefore necessary to use more approximate measures (i.e., appropriate surrogates for character diversity). A good surrogate for character diversity is species or higher taxon richness (Humphries et al. 1995). Generally, measurements at the species level are more precise than higher taxon measures, but gathering data at the species level has much higher costs (Williams 1998). Williams and Gaston (1994) found that family richness is an inexpensive and good predictor for species richness for a variety of groups and regions, whereas Balmford and Gaston (1999) point out that investing in detailed biodiversity inventories ensures efficient and representative reserve selection.

Another possible 'shortcut' surrogate is to use certain species or groups of species as indicators for wholesale biodiversity. For example, Myers (1988) and Myers et al. (2000)

used higher plants as an indicator group in identifying global biodiversity hotspots. Prendergast et al. (1993) investigated birds, butterflies, dragonflies, liverworts and aquatic plants in Britain, Howard et al. (1998) investigated woody plants, large moths, butterflies, birds and small mammals in Uganda and Lovett et al. (2000) investigated groups of plants in Africa; all found poor spatial congruence in the species richness of different taxa.

The appropriateness of using threatened species or 'flagship' species as indicators has also been assessed. Whereas the use of flagship species is no better than a random selection in capturing overall biodiversity (Williams et al. 2000), the use of threatened species as indicators is far better even though they seem to represent wholesale biodiversity relatively inefficiently compared to using all taxa in the area selection analysis (Lund and Rahbek 2000).

The use of indicator groups for setting overall conservation priorities should be done with great caution and must rely on demonstrated spatial congruence between a proposed indicator group in the area in question and the fraction of biodiversity that is to be conserved. Whether congruence is found in a particular case may depend on the area selection method used. Howard et al. (1998) found that species richness of one taxon group poorly captures the diversity of other taxon groups, while sets of priority areas selected for one taxonomic group using complementarity methods captured the species richness of other taxonomic groups fairly efficiently.

10.1.2 AREA SELECTION METHODS: THE RICHNESS APPROACH

One set of methods prioritizes those areas that are particularly rich in species or contain great numbers of endemic or rare species. This 'hotspot' approach was first coined by Myers (1988), who defined a hotspot as an area of extreme endemism that faces extreme threat. Reid (1998) defined the term hotspot more broadly as '...a geographical area that ranks particularly high on one or more axes of species richness, levels of endemism, numbers of rare and threatened species, and intensity of threat'.

However, he failed to distinguish between species diversity and endemism diversity, which may, or may not, be congruent. Choosing areas with the highest numbers of species as priorities for biodiversity conservation has been a popular method (see Williams, 1998, and references herein). For example, WWF and IUCN used this approach to identify centres of plant diversity (WWF and IUCN 1994). Prendergast et al. (1993) discussed the consequences of selecting the top 5% of areas within Britain by species richness as conservation areas. The method has the appeal of dealing with species-occurrence data with apparent quantitative rigor and the further advantage that exact identity of each species is not required for the method to work (Williams 1998).

Richness of rarity or endemism gives greater weight to the more narrowly distributed species. This approach has been used, for example, to identify areas particularly rich in endemic birds (ICPB 1992) or plants (Myers 1988; Myers et al. 2000). While some studies investigate species with range sizes below a certain threshold, others use a continuous function of range size for all species designed to give greater weight to rare species. The former approach has the advantage that data are only required for the rare species, but the use of an arbitrary threshold has been criticized because it will always miss important species with marginally larger ranges than the threshold (Williams 1998). On the other hand, there is no natural formula for weighting range size, so any range-size weighting algorithm is arbitrary (Williams 1998).

10.1.3   COMPLEMENTARITY METHODS

The richness approach has been criticized by many as inefficient in selecting reserves for biodiversity conservation (Vane-Wright et al. 1991; Williams et al. 1996; Csuti et al. 1997; Mace et al. 2000). An alternative approach, based on complementarity, was made popular by Vane-Wright et al. (1991). Using this approach, areas are selected in a stepwise fashion selecting at each step the area most complementary to the existing reserves or the previously selected areas. Areas are thus selected from the database for their contribution to the overall network of reserves rather than for their individual properties (Pressey et al. 1993). Complementarity methods are more efficient than the hotspot or richness approaches in selecting priority areas that fulfill a given biodiversity conservation goal (Vane-Wright et al. 1991; Williams et al. 1996; Csuti et al. 1997). In practice, this means that fewer areas are generally required to protect all species at least once. The efficiency of complementarity methods has been illustrated in empirical studies for identifying important areas for birds in Great Britain (Williams et al. 1996) and for terrestrial vertebrates in Oregon (Csuti et al. 1997).

The choice of area selection algorithm is an important determinant of the efficiency of complementarity methods. Most of the algorithms used are heuristic and able to find close to minimum sets relatively fast, even for large data sets. The disadvantage of heuristic algorithms is that they do not necessarily always find the optimal set of areas (Williams 1998). Underhill (1994) criticized suboptimal heuristic algorithms and proposed to use integer-programming techniques to find optimal solutions. Others have found that such algorithms are too slow for large data sets and therefore inapplicable to practical conservation planning. It has been suggested that 'intelligent' heuristic algorithms provide a middle ground between algorithm optimality and practicality (Pressey et al. 1996).

In most regions, there are many ways to combine sites into a network, and a complementarity algorithm may find several alternative solutions that are equally efficient in representing biodiversity. This is an advantage since it provides the conservation planner with flexibility in the design of networks because areas unsuited for selection can be disregarded without compromising the overall goal (Pressey et al. 1993). Some sites, however, may be irreplaceable because they harbour species that are found nowhere else or are essential for achieving a given conservation goal in another way. Measuring irreplaceability is a fundamental way of measuring the conservation value of any area, and it can be defined as the extent to which the options for conservation are lost if a specific site is lost (Pressey et al. 1993). The *near-minimum set* algorithm was introduced by Margules et al. (1988) and has been implemented in the WORLDMAP software (Williams 2001; http://www.nhm. ac.uk/science/projects/worldmap/). It is an example of a heuristic algorithm that uses the principle of complementarity and addresses flexibility and irreplaceability. The algorithm described in Endnote 1 has been used for area selection by Csuti et al. (1997) and Lund and Rahbek (2000).

10.1.4   NONTAXONOMIC AREA SELECTION METHODS

Selecting reserves based on biological or taxonomical criteria is not the only solution, and historically other selection criteria have been far more widespread. In some countries, many of today's reserves are merely wilderness areas, the land that nobody wanted. This means land that originally was thought to have limited value for agricultural or urban development and therefore was left for other purposes, such as conservation (Pressey 1994). The protection of nature for its own sake has historically been secondary to major extractive uses such as agriculture, mining and forestry, and many support the view that reserves have often been dedicated for what they are not, rather than for what they are (Pressey 1994).

Other reasons for reserve selection, not connected with representing biodiversity, include recreational values and spectacular scenery, potential for tourism, private hunting reserves, water catchment areas and control of soil erosion. Such ad hoc reserve selection causes bias in the representation of biodiversity and increases the cost of achieving a representative reserve system because many reserves may be selected despite their limited value for biodiversity (Pressey 1994).

Another non-taxonomic approach is to use wilderness as a criterion for selecting reserves. This method is favoured by lobby groups in their campaigning for the world's ancient forests (Greenpeace 2002), and members of Global Forest Watch of the World Resources Institute have recently finalized studies in Africa, Indonesia and North America where they identify and map the distribution of large tracts of low-access forests (Forest Watch Indonesia and Global Forest Watch 2002; Minnemeyer 2002; Noguerón 2002). Wilderness areas are important — for example, as storehouses of biodiversity, regulators of the global and regional climates and homelands for indigenous peoples (Mittermeier et al. 1998). However, selecting wilderness areas may not ensure efficient conservation of biodiversity and it may fail to conserve threatened and highly diverse areas (Pressey 1994). Even in cases of very low availability of biological data, crude biological surrogates probably serve as better tools in selection of reserves for biodiversity conservation than does the wilderness criterion (Pressey 1994).

10.2   CONSERVATION OF RATTANS ON BORNEO

With an area of 751,000 km^2, Borneo is the third largest island in the world, stretching from about 4°S to 7°N. Politically, Borneo is divided among three countries: Brunei Darussalam, Indonesia (Kalimantan) and Malaysia (the states of Sabah and Sarawak) (Microsoft 2000; see Figure 10.1). Large tracts of Borneo, particularly the southern and eastern regions, consist of hilly lowlands and swampy plains. The central and north-western regions are dominated by mountain ranges with peaks rising to between 1000 m and the height of Mt. Kinabalu, which, at 4101 m, is the highest peak between the Himalayas and New Guinea. Most of the island consists of geologically young sedimentary rocks, including sandstone, mudstone and limestone (Paine et al. 1985).

The dominant vegetation type is evergreen rain forest of various types, determined primarily by soil types (United Nations Forum on Forests 2002). Tall, lowland and hill forests (up to 1000 m altitude) are dominated by trees of the family Dipterocarpaceae. This family includes most of the commercial timber trees. Montane forest of lower stature replaces dipterocarp forest on hills and mountains above 1000 m (Paine et al. 1985) (Figure 10.2).

The moist forests of the tropics are considered to be the most diverse ecosystems on Earth, consisting of an estimated 60% of all species despite the fact that they cover only approximately 7% of the land surface and 2% of the surface of the globe (Secretariat to the Convention on Biological Diversity 2001a). Among these, the forests of Southeast Asia are known for their high biodiversity, arguably among the greatest in the world (FAO 2001). The island of Borneo is no exception because it contains between 9000 and 15,000 plant species; the uncertainty reflects a large gap in knowledge (Wong 1998). Approximately

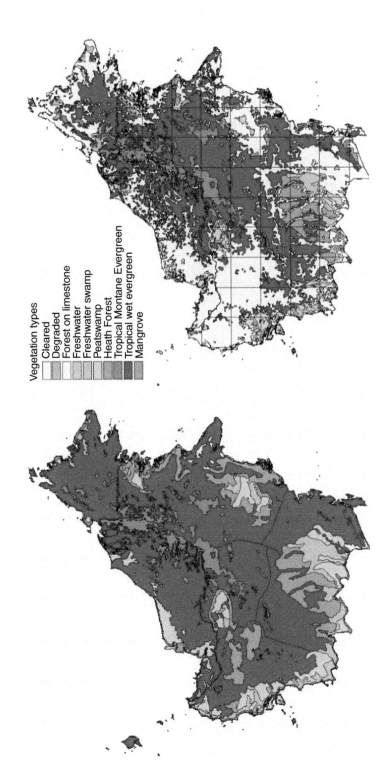

FIGURE 10.1 (**Colour Figure 10.1 follows p. 180.**) The original (left) and current (right) extent of the different vegetation types/land uses represented on Borneo. Based on data from the ASEAN Regional Centre for Biodiversity Conservation. (MacKinnon, pers. comm.)

FIGURE 10.2 **(Colour Figure 10.2 follows p. 180.)** Map showing the protected areas on Borneo. Dark green indicates natural vegetation and bright green cleared or degraded vegetation. Orange polygons are existing reserves; bright blue are proposed reserves. (MacKinnon, pers. comm.)

one third of the plant species are endemic to the island (Myers 1988). The tree flora of the north-western and northern parts of Borneo appears richer than the remaining parts. Of the 311 species of tree recorded from Borneo, 108 species are endemic. Most of these endemics are found in the north-western and northern parts (Soepadmo and Wong 1995). These trends were earlier explained by lower collection intensity in Kalimantan compared to Sabah, Sarawak and Brunei, but further work has confirmed these trends (Wong 1998). Mt. Kinabalu in Sabah is a major centre of plant diversity with some 4500 plant species recorded from this area alone. It is the richest botanical site on Borneo and in all of Asia west of New Guinea (Wong 1998).

10.2.1 Land Use Changes

Borneo is part of the Sundaland hotspot identified by Myers et al. (2000), who indicated that the area experiences exceptional habitat loss. The Sundaland hotspot includes Peninsular Malaysia, Java, Sumatra and Borneo. Myers and colleagues estimate that 7.8% of the natural vegetation remains in this area.

Due to the division of Borneo into four political units, data are rarely produced covering the whole island. FAO (2001) estimates that both Malaysia and Indonesia lose approximately 1.2% of their forest areas annually, but this figure also includes records of deforestation outside Borneo. In comparison to Malaysia and Indonesia, Brunei currently has a relatively stable forest cover (FAO 2001). The causes of deforestation and forest degradation are complex. Among the main factors are conversion of forest into plantations and agriculture, unsustainable shifting cultivation, unsustainable logging practices, illegal logging, forest

TABLE 10.1
Lost Natural Vegetation

Brunei	2,051.06	–35%
Sabah	40,767.02	–54%
Sarawak	57,316.32	–46%
Kalimatan	222,112.59	–41%
Borneo	322,246.99	–43%

TABLE 10.2
Extent of Reserves in the Four Political
Units on Borneo and for Borneo in Total

Sabah	4,576.7	6.1%
Brunei	997.9	17.0%
Sarawak	3,720.2	3.0%
Kalimantan	49,962.1	9.2%
Borneo	**59,256.8**	**7.89%**

Source: MacKinnon, personal communication.

fires and encroachment (Paine et al. 1985; Blakeney 2001; FAO 2001; Forest Watch Indonesia and Global Forest Watch 2002).

Using data compiled by the ASEAN Regional Centre for Biodiversity Conservation
(MacKinnon, pers. comm.; see Methods for further details), we have estimated the total loss
of natural vegetation on Borneo (see Table 10.1, Table 10.2, Figure 10.2 and Endnote 1). The
total loss of natural vegetation on Borneo vegetation is an estimated 322,247 km^2 or 43%
of the total land area. The changes have been most dramatic in Sabah and Sarawak (loss of
54 and 46% of the natural vegetation, respectively) but also very significant in Kalimantan
and Brunei (loss of 41 and 35% of the natural vegetation, respectively). Overall, the vegetation that is most diminished is tropical wet evergreen forest. Proportionally, mangrove is
the vegetation type that has been diminished most (–72%), followed by freshwater swamps
(–53%), heath forests (–52%) and peat swamps (–51%).

GLOBIO (global methodology for mapping human impact on the biosphere, http://www.
globio.info) has developed a future scenario for the consequences of human impact on biodiversity. The scenario is based on data describing existing infrastructure, historic growth
rates of infrastructure, availability of petroleum and mineral reserves, vegetation cover,
population density, distance to coast and projected development. The outcome is a simple
overview of possible future human impacts on biodiversity, assuming continued growth in
demand for natural resources and the associated infrastructure development (Groombridge
and Jenkins 2002). Following this scenario, losses of biodiversity are likely to be particularly severe in Southeast Asia (together with the Congo Basin and parts of the Amazon) and
the GLOBIO scenario forecast continued large-scale conversions of natural habitat and that
no primary forests will be left on Borneo in 2030 (http://www.globio.info/region/asia/).

10.2.2 Reserves

Approximately 70 reserves are found on Borneo, covering almost 60,000 km^2 or 8% of the land area (MacKinnon, pers. comm.). In addition to the existing reserves are a great number of proposed reserves that, if implemented, would double the reserves area of Borneo.

Brunei has protected as much as 17.0% of its land area as reserves despite its relatively small size, whereas Sarawak only protects 3.0% of its land as reserves. Sabah and Kalimantan are intermediate with 6.1 and 9.2%, respectively (see Table 10.3 and Figure 10.1 for details). Management and enforcement of protection measures in reserves as well as the level of protection outside reserves may have serious implications for the effective protection of biodiversity on Borneo, but addressing these issues is beyond the scope of this study.

The tables are based on unpublished GIS data from the ASEAN Regional Centre for Biodiversity Conservation (MacKinnon, pers. comm.). The data originate from the World Conservation and Monitoring Centre and have been augmented by MacKinnon using vegetation maps, land cover maps, geological maps and protected area boundary maps (MacKinnon, pers. comm.). 'Original vegetation' is here to be understood as the natural vegetation that would be found in Borneo if man had not degraded it.

10.3 RATTANS

Rattans are spiny, climbing palms (Arecaceae) that belong to the subfamily Calamoideae (Asmussen et al. 2000). Thirteen of the 22 genera in that subfamily are classified as rattan genera. Worldwide, there is a total of 560–600 species, and two thirds of these (374) belong to the genus *Calamus* (Govaerts and Dransfield 2005). Whereas Calamoideae forms a monophyletic group within Arecaceae (Asmussen et al. 2000), the climbing habit of the rattans seems to have several origins within the subfamily (Baker and Dransfield 2000). Rattans grow single stemmed or form multistemmed large clusters in one individual. Most species do not branch from the aerial stems; the genera *Korthalsia* and *Laccosperma* are the exceptions. Leaves are produced sequentially, one at the time, and consist of a tubular sheath that narrows into a petiole in the upper end and continues into the leaflet-bearing rachis.

The leaf blade and the leaf sheath often have spines. The rachis often extends beyond the terminal leaflets into a whip-like climbing organ, a cirrus. Climbing organs can also develop from sterile inflorescences in the form of a flagellum borne on the leaf sheath near the node. Flagellae are unique to *Calamus* but not all species of *Calamus* bear flagellae. The flagellum and the cirrus bear short, reflexed spines. The inflorescences are produced singly at the nodes and vary greatly in size and structure. The flower is hermaphroditic, or unisexual, and either monoecious (only in *Oncocalamus*) or dioecious. The rattan fruit is characteristically covered with vertical rows of overlapping reflexed scales (Uhl and Dransfield 1987). The names and authors of the taxa used in the present analyses are in Table 10.4.

10.3.1 Distribution and Ecology

All rattans are found in the Old World tropics and subtropics, ranging from equatorial Africa, the Indian subcontinent, Sri Lanka, the foothills of the Himalayas, southern China through the Malay Archipelago to Australia and the western Pacific as far as Fiji

TABLE 10.3

Changes in Natural Vegetation on Borneo

Current vegetation

	Brunei Area	%	Sabah Area	%	Sarawak Area	%	Kalimantan Area	%	Borneo Area	%
Cleared degraded	1188.5	20.24	34739.1	46.29	57321.7	45.69	222078.8	40.78	315328.1	41.99
freshwater swamp	787.3	13.41	6024.5	8.03	0.0	0.00	0.0	0.00	6811.9	0.91
heath forest	77.0	1.31	832.0	1.11	187.0	0.15	18283.1	3.36	19379.2	2.58
limestone	0.0	0.00	94.7	0.13	652.2	0.52	37584.7	6.90	38331.7	5.10
mangrove	0.00	0.00	23.4	0.03	75.9	0.06	1710.3	0.31	1809.6	0.24
peat swamp	202.6	3.45	699.7	0.93	972.3	0.78	5882.6	1.08	7757.2	1.03
tropical montane	1262.9	21.50	38.6	0.05	5717.3	4.56	22204.1	4.08	29222.9	3.89
evergreen	74.6	1.27	3071.4	4.09	14762.6	11.77	27951.3	5.13	45859.9	6.11
tropical wet evergreen	2280.4	38.83	29408.5	39.19	45762.5	36.48	208915.5	38.36	286366.9	38.13
not classified (sq. km)	0.0	0.00	118.3	0.16	0.0	0.00	14.4	0.00	132.7	0.02
current vegetation Total area	5873.2	100.00	75050.3	100.00	125451.6	100.00	544624.9	100.00	751000.0	100.00

Original vegetation

	Brunei Area	%	Sabah Area	%	Sarawak Area	%	Kalimantan Area	%	Borneo Area	%
Freshwater swamp	93.0	1.56	2601.3	3.47	299.5	0.24	38345.0	7.04	41338.7	5.50
heath forest	88.0	1.48	278.7	0.37	1381.7	1.10	78454.4	14.40	80202.8	10.68
limestone	0.0	0.00	106.6	0.14	384.9	0.31	2075.5	0.38	2567.1	0.34
mangrove	252.4	4.24	4712.8	6.28	3186.3	2.54	19515.1	3.58	27666.6	3.68
peat swamp	1675.5	28.17	314.0	0.42	13620.5	10.86	43535.5	7.99	59145.5	7.87
tropical montane	74.6	1.25	3500.7	4.66	16899.5	13.47	28848.2	5.30	49323.0	6.57
tropical wet evergreen	3765.1	63.29	63539.5	84.66	89663.0	71.48	333884.9	61.30	490852.4	65.35
not classified (sq. km)	0.0	0.00	0.1	0.00	10.8	0.01	0.0	0.00	10.9	0.00
original vegetation Total area	5948.5	100.00	75053.7	100.00	125446.2	100.00	544658.7	100.00	751107.1	100.00

TABLE 10.3 (continued)
Changes in Natural Vegetation on Borneo

Vegetation type	Brunei Area %	Sabah Area %	Sarawak Area %	Kalimantan Area %	Borneo Area %
Freshwater swamp	-16.0 -17.18 -88.0	-1769.3 -68.01	-112.4 -37.55	-20061.9 -52.32	-21959.6 -41871.1 / -53.12 -52.21 -29.51
heath forest	-100.00 0.0 -49.8	-184.0 -66.02	-729.4 -52.79	-40869.7 -52.09	-757.5 -19909.4 / -71.96 -50.59 -7.02
limestone mangrove	-19.74 -412.6	-83.2 -78.04	-309.0 -80.27	-365.3 -17.60	-29922.6 -3463.2 / -41.66 0.02 %
peat swamp tropical	-24.63 0.0 -0.02	-4013.1 -85.15	-2214.0 -69.48	-13632.5 -69.86	-204485.5 121.8 / -42.91
montane evergreen	-1484.7 -39.43 0.0	-275.4 -87.70	-7903.2 -58.02	-21331.4 -49.00	Borneo Area
tropical wet	0.00	-429.3 -12.26	-2136.9 -12.64	-896.9 -3.11	-322246.99
evergreen not	-2051.06 -34.92	-34131.0 -53.72	-43900.5 -48.96	-124969.3 -37.43	
classified (sq. km)	118.2 0.16	-10.8 -0.01	14.4 0.00		
Change in natural vegetation	-40767.02 -54.32	-57316.32 -45.69		-222112.59 -40.78	
Total change					

Note: The table can be expanded at will.

TABLE 10.4

Rattans of Borneo and Their Distributions

Taxon	Distributions
Calamus L.	
Calamus acanthochlamys J. Dransf.	Local endemic (32)
Calamus acuminatus Becc.	Widespread endemic (101)
Calamus amplijugus J. Dransf.	Widespread endemic (145)
Calamus ashtonii J. Dransf.	Widespread endemic (42)
Calamus axillaris Becc.	Narrow (5); also in Thailand, Peninsular Malaysia
Calamus bacularis Becc.	Widespread endemic (85)
Calamus blumei Becc.	Very widespread (990); also in Sumatra, Thailand, and Peninsular Malaysia
Calamus caesius Bl.	Very widespread (1086); also in Sumatra, Thailand, Peninsular Malaysia, and Palawan
Calamus comptus J. Dransf.	Widespread endemic (65)
Calamus congestiflorus J. Dransf.	Local endemic (12)
Calamus conirostris Becc.	Very widespread (994); also in Peninsular Malaysia and Sumatra
Calamus conjugatus Furtado	Narrow endemic (2)
Calamus convallium J. Dransf.	Very widespread endemic (425)
Calamus corrugatus Becc.	Widespread endemic (68)
Calamus crassifolius J. Dransf.	Narrow endemic (2)
Calamus diepenhorstii var. *diepenhorstii* Miq.	Widespread (40); also in Sumatra, Thailand, and Peninsular Malaysia
Calamus diepenhorstii var. *major* J. Dransf.	Narrow endemic (3)
Calamus divaricatus Becc.	Very widespread endemic (341)
Calamus elopurensis J. Dransf.	Widespread endemic (98)
Calamus erinaceus (Becc.) J. Dransf.	Widespread (88); also in Thailand, Sumatra, Peninsular Malaysia, and Palawan
Calamus erioacanthus Becc.	Widespread endemic (41)
Calamus fimbriatus Valkenburg	Narrow endemic (1)
Calamus flabellatus Becc.	Very widespread (993); also in Sumatra and Peninsular Malaysia
Calamus gibbsianus Becc.	Local endemic (26)
Calamus gonospermus Becc.	Widespread endemic (112)

TABLE 10.4 (continued)
Rattans of Borneo and Their Distributions

Taxon	Distributions
82	
Calamus hepburnii J. Dransf.	Narrow endemic (1)
Calamus hispidulus Becc.	Very widespread endemic (459)
Calamus hypertrichosus Becc.	Narrow endemic (1)
Calamus impar Becc.	Narrow endemic (1)
Calamus javensis Bl.	Very widespread (999); also in Thailand, Sumatra, Peninsular Malaysia, Palawan and Java
Calamus kiahii Furtado	Widespread endemic (61)
Calamus laevigatus var. *laevigatus* Mart.	Very widespread (575); also in Sumatra and Peninsular Malaysia
Calamus laevigatus var. *mucronatus* (Becc.) J. Dransf.	Very widespread endemic (559)
Calamus laevigatus var. *serpentinus* J. Dransf.	Narrow endemic (6)
Calamus lambirensis J. Dransf.	Local endemic (19)
Calamus leloi J. Dransf.	Widespread endemic (134)
Calamus lobbianus Becc.	Widespread (63); also in Peninsular Malaysia
Calamus maiadum J. Dransf.	Narrow endemic (4)
Calamus malawaliensis J. Dransf.	Narrow (1); also in Palawan
Calamus manan Miq.	Widespread (55); also in Thailand, Sumatra, and Peninsular Malaysia
Calamus marginatus (Bl.) Mart.	Very widespread (717); also in Sumatra
Calamus mattanensis Becc.	Widespread endemic (261)
Calamus mesilauensis J. Dransf.	Narrow endemic (3)
Calamus microsphaerion Becc. Vel Valde	Narrow (2); also in the Philippines
Calamus muricatus Becc.	Very widespread endemic (670)
Calamus myriacanthus Becc.	Widespread endemic (216)
Calamus nanodendron J. Dransf.	Local endemic (18)
Calamus nematospadix Becc.	Widespread endemic (69)
Calamus nielsenii J. Dransf.	Narrow endemic (1)
Calamus nigricans Valkenburg	Narrow endemic (1)
Calamus optimus Becc.	Very widespread endemic (1087)
Calamus ornatus Bl.	Very widespread (1086); also in Sumatra, Peninsular Malaysia, Thailand, Java, Sulawesi, and the Philippines

Calamus oxleyanus Teijsm. & Binn. ex Miq.	Local (16); also in Thailand, Sumatra, and Peninsular Malaysia
Calamus pandanosmus Furtado	Local (14); also in Sumatra, Southern Thailand, and Peninsular Malaysia

83

Calamus paspalanthus Becc.	Very widespread (651); also in Sumatra and Peninsular Malaysia
Calamus paulii J. Dransf.	Narrow endemic (1)
Calamus pilosellus Becc.	Very widespread endemic (1087)
Calamus poensis Becc.	Narrow endemic (2)
Calamus pogonacanthus Becc.	Very widespread endemic (597)
Calamus praetermissus J. Dransf.	Widespread endemic (287)
Calamus pseudoulur Becc.	Widespread endemic (310)
Calamus psilocladus J. Dransf.	Narrow endemic (1)
Calamus pygmaeus Becc.	Narrow endemic (7)
Calamus rhytidomus Becc.	Widespread endemic (260)
Calamus ravidus Becc.	Widespread endemic (42)
Calamus sabalensis J. Dransf.	Narrow endemic (2)
Calamus sabensis Becc.	Narrow endemic (1)
Calamus sarawakensis Becc.	Very widespread endemic (449)
Calamus schistoacanthus Bl.	Narrow endemic (2)
Calamus scipionum Lour.	Very widespread (1087); also in Indochina, Thailand, Peninsular Malaysia, Sumatra and Palawan
Calamus semoi Becc.	Local endemic (18)
Calamus sordidus J. Dransf.	Widespread endemic (91)
Calamus spectatissimus Furtado	Narrow (4); also in Sumatra, Southern Thailand and Peninsular Malaysia
Calamus subinermis H. Wendl.	Local; also in the Philippines and Sulawesi (13)
Calamus tapa Becc.	Narrow endemic (5)
Calamus temburongii J. Dransf	Narrow endemic (2)
Calamus tenompokensis Furtado	Local endemic (22)
Calamus tomentosus Becc.	Narrow (1); also in Peninsular Malaysia
Calamus trachycoleus Becc.	Widespread endemic (102)
Calamus usitatus Blanco	Local (15); also in the Philippines
Calamus winklerianus Becc.	Narrow endemic (1)
Calamus zonatus Becc.	Very widespread endemic (331)

TABLE 10.4 (continued)
Rattans of Borneo and Their Distributions

Taxon	Distributions
***Ceratolobus* Bl.**	
Ceratolobus concolor Bl.	Very widespread (439); also in Sumatra
Ceratolobus discolor Becc.	Widespread (318); also in Sumatra
84	
Ceratolobus subangulatus (Miq.) Becc.	Very widespread (653); also in Sumatra, Peninsular Malaysia and Southern Thailand
***Daemonorops* Bl.**	
Daemonorops acamptostachys Becc.	Widespread endemic (67)
Daemonorops asteracantha Becc.	Local endemic (16)
Daemonorops atra J. Dransf.	Widespread endemic (307)
Daemonorops banggiensis J. Dransf.	Narrow endemic (1)
Daemonorops collarifera Becc.	Widespread endemic (162)
Daemonorops crinita Bl.	Widespread (128); also in Sumatra
Daemonorops cristata Becc.	Widespread endemic (224)
Daemonorops didymophylla Becc.	Very widespread (991); also in Thailand, Sumatra and Peninsular Malaysia
Daemonorops draco (Willd.) Bl.	Narrow (1); also in Sumatra
Daemonorops elongata Bl.	Widespread endemic (318)
Daemonorops fissa Bl.	Very widespread endemic (1087)
Daemonorops formicaria Becc.	Widespread endemic (236)
Daemonorops hallieriana Becc.	Narrow endemic (2)
Daemonorops hystrix (Willd.) Bl.	Very widespread endemic (1087)
Daemonorops ingens J. Dransf.	Widespread endemic (241)
Daemonorops korthalsii Bl.	Very widespread endemic (991)
Daemonorops longipes (Griff.) Mart.	Very widespread (601); also in Sumatra and Peninsular Malaysia
Daemonorops longispatha Becc.	Local endemic (25)
Daemonorops longistipes Burret	Widespread endemic (90)
Daemonorops maculata J. Dransf.	Widespread endemic (95)
Daemonorops micracantha (Griff.) Becc.	Very widespread (997); also in Peninsular Malaysia

Daemonorops microstachys Becc. — Very widespread endemic (586)

Daemonorops oblata J. Dransf. — Widespread endemic (40)

Daemonorops oxycarpa Becc. — Widespread endemic (196)

Daemonorops periacantha Miq — Very widespread (396); also in Sumatra and Peninsular Malaysia

Daemonorops pumila Valkenburg — Narrow endemic (1)

Daemonorops ruptilis var. *acaulescens* J. Dransf. — Local endemic (9)

Daemonorops ruptilis var. *ruptilis* Becc. — Widespread endemic (191)

Daemonorops sabut Becc. — Very widespread (991); also in Peninsular Malaysia

85

Daemonorops scapigera Becc. — Local (23); also in Peninsular Malaysia and Natuna

Daemonorops serpentina J. Dransf. — Narrow endemic (6)

Daemonorops sparsiflora Becc. — Very widespread endemic (512)

Daemonorops spectabilis Becc. — Widespread endemic (192)

Daemonorops unijuga J. Dransf. — Narrow endemic (1)

Korthalsia Bl.

Korthalsia angustifolia Bl. — Widespread endemic (37)

Korthalsia cheb Becc. — Very widespread endemic (336)

Korthalsia concolor Burret — Local endemic (23)

Korthalsia debilis Bl. — Local; also in Sumatra (10)

Korthalsia echinometra Becc. — Very widespread (996); also in Peninsular Malaysia and Sumatra

Korthalsia ferox Becc. — Widespread endemic (177)

Korthalsia flagellaris Miq. — Local; also in Sumatra, Peninsular Malaysia and Southern Thailand (10)

Korthalsia furcata Becc. — Local endemic (13)

Korthalsia furtadoana J. Dransf. — Widespread endemic (149)

Korthalsia hispida Becc. — Widespread (215); also in Sumatra and Peninsular Malaysia

Korthalsia jala J. Dransf. — Widespread endemic (137)

Korthalsia paucijuga Becc. — Narrow endemic (3)

Korthalsia rigida Bl. — Very widespread (1086); also in Thailand, Sumatra, Peninsular Malaysia and Palawan

Korthalsia robusta Bl. — Widespread (258); also in Sumatra

Korthalsia rostrata Bl. — Very widespread (1087); also in Thailand, Sumatra and Peninsular Malaysia

TABLE 10.4 (continued)
Rattans of Borneo and Their Distributions

Taxon	Distributions
***Plectocomia* Mart. ex Bl.**	
Plectocomia elongata Mart. ex Bl.	Narrow (4); also in Peninsular Malaysia, Sumatra, Java, Southern Thailand and Palawan
Plectocomia mulleri Bl.	Very widespread (993); also in Peninsular Malaysia
Plectocomia pygmaea Madulid	Narrow endemic (1)
***Plectocomiopsis* Becc.**	
Plectocomiopsis geminiflora (Griff.) Becc.	Very widespread (1087); also in Peninsular Malaysia, Sumatra, Thailand and Laos
Plectocomiopsis mira J. Dransf.	Very widespread (993); also in Peninsular Malaysia and Sumatra
Plectocomiopsis triquetra (Becc.) J. Dransf.	Widespread endemic (119)
86	
***Pogonotium* J. Dransf.**	
Pogonotium divaricatum J. Dransf.	Local endemic (28)
Pogonotium moorei J. Dransf.	Narrow endemic (1)
Pogonotium ursinum (Becc.) J. Dransf.	Local (13); also in Peninsular Malaysia
***Retispatha* J. Dransf.**	
Retispatha dumetosa J. Dransf.	Widespread endemic (131)

Notes: Extent of occurrence is measured as the number of grid cells in which the taxon occurs according to its estimated distribution. Summary statistics: genera: 8; species: 140; taxa (species and varieties): 144; endemic species: 95; endemic taxa: 98.

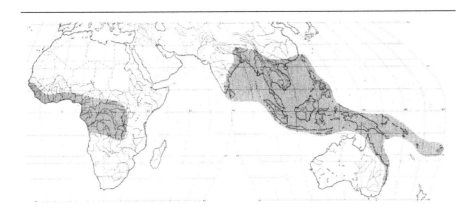

FIGURE 10.3   The distribution of rattans worldwide. (Based on Uhl, N.W. and Dransfield, J. 1987. *Genera Palmarum*, Allen Press, Lawrence, KS.)

(Figure 10.3). The greatest diversity of genera and species are found in Southeast Asia (Dransfield and Manokaran 1993). The large diversity of rattan species and their wide geographical range are matched by a great ecological diversity, which makes generalizations about their ecology difficult. Within their natural distribution, rattan species can be found in the majority of forest types and on most soil and rock types. A range of rattan species live in different soil moisture regimes, from swamps to dry ridge tops. Light intensity is also an important factor determining the distribution of rattans within the forests where they occur (Dransfield and Manokaran 1993).

10.3.2   Uses

Because of their strength, flexibility and uniformity, the bare stems of rattans are very widely used for various construction purposes (Dransfield and Manokaran 1993). In rural areas, rattans collected from wild populations are widely used for a range of purposes, such as baskets, mats, furniture, fish and animal traps, bird cages, ropes and fences (Dransfield and Manokaran 1993; Andersen et al. 2000). In an extensive study of two indigenous communities in Sarawak, Christensen (2002) found that the communities use 38 species of rattan and rattans are their most important source of fibres. Andersen et al. (2001, in press) found that rattans are also of great importance to local people in Sabah for numerous purposes. It is likely that these patterns in rural rattan use are reflected in most communities on Borneo that practice a more or less traditional lifestyle.

Perhaps as much as 20% of the rattan species are used commercially as whole stems, especially for furniture frames, or as splits, peels and cores for matting and basketry. Though rattan cultivation is increasing, the commercial use of rattans relies mainly on collection of rattans from wild populations (Dransfield and Manokaran 1993). The incomes from selling rattan canes are an important source of funds for many local communities (Kiew and Pearce 1990), and rattans are generally regarded as an important non-timber forest product (NTFP) in Sabah and Sarawak (Anonymous 1985).

10.3.3   Rattans on Borneo

Rattans are one of the few relatively well-studied plant groups on Borneo. Their taxonomy is well described, and their diversity has been documented in manuals for Sabah, Sarawak and Brunei (Dransfield 1984a, 1992, 1997; Kirkup et al. 1999). An interactive manual for Borneo (Dransfield and Patel 2005) has been published. Furthermore, a number of studies have documented the rattan floras of specific areas of Borneo, including Gunung Mulu National Park (Dransfield 1984b), Kubah National Park (Pearce 1994), Maliau Basin (Gait et al. 1998) and Tabin Wildlife Reserve (Boje 2000; Andersen et al. 2001, in press). These studies indicate that Borneo harbours a rich rattan flora with a high level of endemism. The broader distribution patterns of rattans on Borneo are provided by Dransfield and Patel (2005). These authors have summarized the conservation status of rattans on Borneo as a whole, and studies addressing the conservation status of palms in Malaysia, Sabah and Sarawak provide valuable background information.

Dransfield and Johnson (1990) stress that the conservation status of rattans in Sabah is very sparsely known and identify only two species as threatened. Based on data from Sarawak Forest Department and the literature, Pearce (1989) identifies a total of 44 rattan taxa as threatened in Sarawak (vulnerable or endangered). No detailed studies on the conservation status of rattans in Brunei or Kalimantan have been made. Kiew and Pearce (1990) stress that only half of the palm species in Malaysia are found in protected areas and that these species mainly are widespread. They note that rare and endangered species with narrow distributions are most often not protected, and they highlight the need for a network of protected areas of high endemism.

The main threats to rattans are habitat conversion and degradation through logging and shifting cultivation, and overexploitation of wild populations of valuable rattan species (Pearce 1989; Kiew and Pearce 1990). Some species are most probably overexploited in Sabah and Sarawak due to an increasing demand for raw materials for making furniture and other commercial products (Anonymous 1985; Kiew and Pearce 1990). Even rattans that occur in protected areas may be threatened due to illegal commercial collection, and some forests are clear of certain species (Pearce 1989). Continued overexploitation of rattans and large-scale clear felling of forests may result in extinction of endemic species with narrow distribution (Kiew and Pearce 1990). The gene bank of important commercial rattan species may also be depleted and this may threaten continued commercial rattan exploitation (Kiew and Pearce 1990).

10.3.4 MATERIALS: RATTAN DATA

The data on rattan distribution are from two different sources: existing data from selected herbaria and new collections made by Andersen in Sabah. Specimen data of rattans collected on Borneo and located in the following herbaria were examined (abbreviations according to Holmgren et al. 1990): AAU, BO, C, K, KEP, L, SAN and Sabah Parks herbarium, Sabah, Malaysia. The data were obtained from the herbaria within the period September 2000 to October 2001. A database of rattans collected at Mt. Kinabalu was also examined (Beaman, pers. comm.) as well as literature records of rattans included in Pearce (1994). All data were compiled in a database and were thoroughly examined to correct erroneous taxonomy and discard specimens with dubious identifications. Because data from several herbaria were included, many specimens were duplicates, and some of these duplicates had different identifications. Only one of the duplicates was included in the analysis and in cases of discrepancy between duplicates, the specimen with the most reliable and detailed identity or description was chosen. The criteria for choosing among

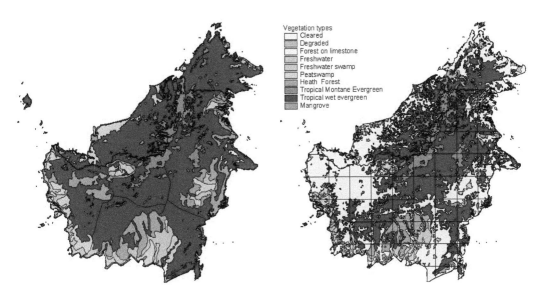

COLOUR FIGURE 10.1 The original (left) and current (right) extent of the different vegetation types/land uses represented on Borneo. Based on data from the ASEAN Regional Centre for Biodiversity Conservation. (MacKinnon, pers. comm.)

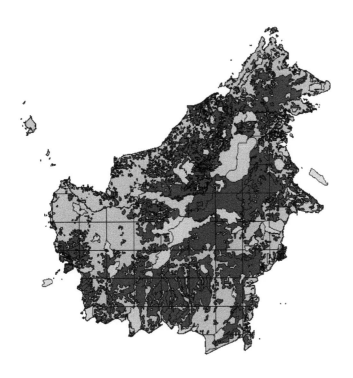

COLOUR FIGURE 10.2 Map showing the protected areas on Borneo. Dark green indicates natural vegetation and bright green cleared or degraded vegetation. Orange polygons are existing reserves; bright blue are proposed reserves. (MacKinnon, pers. comm.)

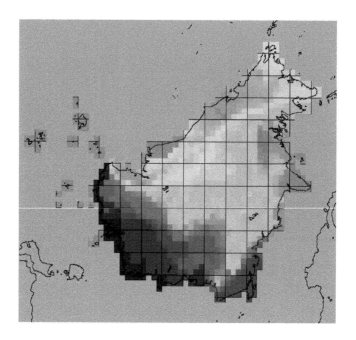

COLOUR FIGURE 10.5 WORLDMAP illustration showing the taxon richness of rattans on Borneo based on estimated distributions. The colours of the grid cells indicate their species richness. Saturated red colours indicate many species and dark blue colours indicate few species.

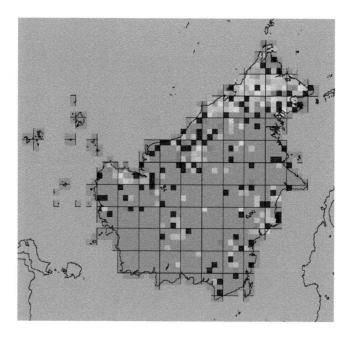

COLOUR FIGURE 10.6 WORLDMAP illustration showing the richness of rattans on Borneo based on confirmed records only. The colours of the grid cells indicate their species richness. Saturated red colours indicate many species and dark blue colours indicate few species.

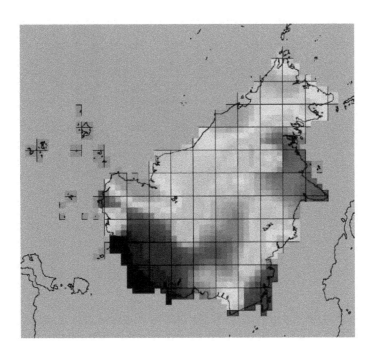

COLOUR FIGURE 10.7 WORLDMAP map showing range-size rarity patterns of rattans on Borneo based on estimated distributions. Saturated red colours indicate high range-size rarity scores and the dark blue colours indicate low values.

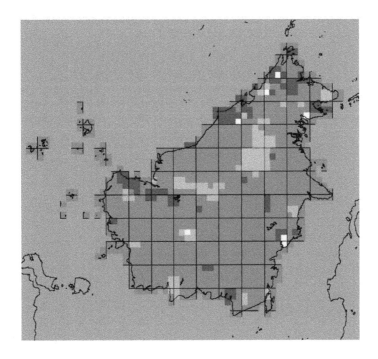

COLOUR FIGURE 10.9 WORLDMAP between 91 grid cells in which 30% is covered by protected areas (green colour) and grid cells selected in a near-maximum coverage for 91 grid cells (blue colour). White cells indicate overlap.

COLOUR FIGURE 10.10 A map showing the locality of the 26 priority areas for rattan conservation on Borneo. Red quadrates are grid cells selected in the preliminary set of priority areas for rattan conservation. Yellow quadrates are additional grid cells selected in the *near-minimum set* analysis in which grid cells with less than 10% natural vegetation were excluded. Together, red and yellow quadrates make up the final set of priority areas for rattan conservation on Borneo. Selected grid cells with a white dot are well protected. Orange polygons are reserves and bright blue polygons are proposed reserves. Dark green illustrates natural vegetation and bright green illustrates cleared or degraded land.

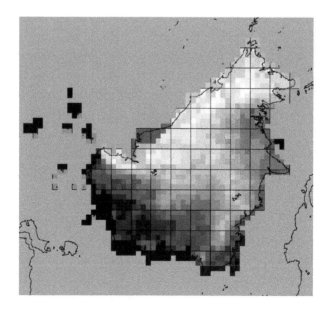

COLOUR FIGURE 10.12 The congruence in richness patterns of rattans and birds and overlay of the richness maps of butterflies and rattans. White colours indicate grid cells of high rattan and bird richness. Green colours indicate that rattan diversity is relatively higher than butterfly diversity. Blue colours illustrate that bird diversity is relatively higher than rattan diversity.

TABLE 10.5

Collection Localities, Collecting Period and Number of Days Spent Collecting, and Forest Types and Altitudes Where Rattans Were Collected

Locality	Collecting period	No. of collecting days	Forest types	Altitude (ma.s.l.)
Crocker Range	October 1999 and September 2000	8	Primary and secondary dipterocarp forest	341–1046
Maliau Basin	October 2000	4	Primary dipterocarp and heath forest	571–1045
Tabin Wildlife Reserve	October 2000	9	Mangrove and secondary dipterocarp forest, including areas with limestone outcrops	40–150

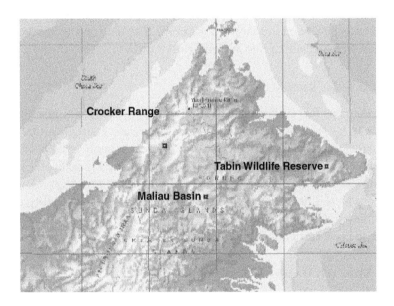

FIGURE 10.4 A map of northern Borneo indicating the geographical location of the three main localities.

duplicates were primarily judgement of the level of rattan expertise demonstrated by the people having determined the duplicates and, secondarily, judgement of the overall level of rattan expertise at the institutions holding the duplicates.

The herbarium collections were improved by rattans collected in three areas in Sabah, Malaysia: Crocker Range, Maliau Basin and Tabin Wildlife Reserve (see Table 10.5 and Figure 10.4 for details on collection localities). Specimens were collected, described and prepared using the methods outlined by Dransfield (1986). The exact geographical position was recorded using GPS (Garmin GPS 12). Andersen identified the specimens with assistance from Diwol Sundaling (Forestry Department, Sabah) and Dransfield. Some commonly occurring, easily recognized rattans were recorded only and not collected. During the field-work on Borneo, a total of 107 rattan specimens were collected or recorded. A complete set of specimens was deposited at SAN. Duplicates of specimens collected in Crocker Range National Park are deposited in Sabah Parks herbarium. Additional duplicates of most specimens are at AAU, C, K, KEP, L, and UKMS. The total data set consists of 5045 specimens. From these, 4829 (96%) had sufficiently precise locality description to allow geographical referencing. The oldest rattan specimen used was collected in July 1803 and the most recent were Andersen's, collected in October 2000. Most of the specimens (80%) were collected after 1970 and almost half of the specimens (46%) were collected after 1990.

10.3.5 OTHER DATA

Detailed data on geographical location and status of protected areas, as well as original and current distribution of vegetation types and land uses were kindly provided in ArcView 3.1 GIS format by John MacKinnon (ASEAN Regional Centre for Biodiversity Conservation). The GIS maps are based on data from the World Conservation and Monitoring Centre and augmented by MacKinnon using vegetation maps, land cover maps, geological maps and protected area boundary maps from the countries concerned (MacKinnon, pers. comm). Part of the data was published in MacKinnon (1997). The data on vegetation and protected areas on Borneo were converted into WORLDMAP format.

10.3.6 GEOGRAPHICAL REFERENCING

Most of the specimens lacked precise geographical coordinates. Because this is a prerequisite for analysis by WORLDMAP and due to the great variation in the quality of the specimens' locality description, a method was developed to apply reliable and precise geographical reference to as many specimens as possible. A grid projection is used in this study; the main purpose of the geographical referencing for any specimen was to locate the grid cell in which it was collected. According to the kind and quality of the locality description on any given specimen, different tools and methods were used to geographically reference the specimen (Endnote 1).

10.3.7 DOT MAPS AND DISTRIBUTION MAPS

The geographically referenced specimens were used to construct 'dot maps' of known collections for all rattan taxa on Borneo. The dot maps were then used to estimate the potential geographical distribution of the taxa. In deciding on the distribution maps, we were assisted by maps of Borneo on rainfall, altitude and Borneo's ecological regions. Justin Moat of the GIS unit at the Royal Botanic Gardens, Kew, provided these maps. The methods and principles used in making the estimated distribution maps included:

- In a case of narrow geographical scatter of the known collections, a line was drawn around the collection localities (dots) to indicate the distribution of the taxon. Narrow geographical scatter was in this context defined as no more than five grid cells between any of the dots.
- In a case of many collections with widespread occurrence and large geographical scatter (practically covering the whole of Borneo), the potential distribution of the taxon was estimated to be the whole of Borneo. Possible unsuitable habitats (such as peat swamps for some taxa) were excluded when these covered whole grid cells.

- In some cases, a taxon was estimated to potentially occur in large areas where it had never been collected. Generally, this was done whenever the distance between two known collections was smaller than 20 grid cells, and one or more of the following factors could qualify the choice: the area was poorly collected, the habitat was supposed to be suitable throughout the intervening area between known collections or the taxon was known to be very rare, even in its locality.
- In other cases, the estimated distribution of a taxon was divided into separate areas of potential distribution. Generally, this was done whenever the distance between two known collections was larger than 20 grid cells. In cases where the potential distribution was split despite shorter gaps, one or more of the following factors qualified the split: lack of knowledge of the habitat (conservative choice), the habitat intervening habitat between two known collections was supposed not to be suitable for the taxon or the taxon was supposed not to occur in the intervening area between two collections because extensive collection efforts in the area had no new discoveries.
- In some cases, the line drawn between two known collections was curved in order to limit the estimated distribution area because the habitat was considered unsuitable for the taxon.
- In many cases, the estimated distribution of a taxon was extrapolated beyond known collections to the nearest coastline, when the distance to the coastline did not exceed five grid cells. For some taxa, known to have a very wide distribution on Borneo, extrapolation of the estimated distribution went somewhat beyond five grid cells. Examples of estimated rattan distribution maps are found in Figure 10.5.

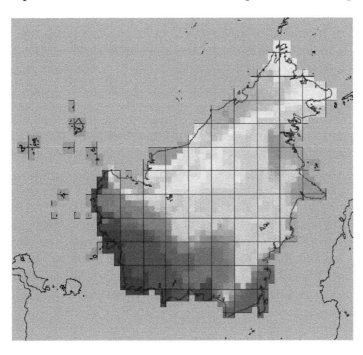

FIGURE 10.5 **(Colour Figure 10.5 follows p. 180.)** WORLDMAP illustration showing the taxon richness of rattans on Borneo based on estimated distributions. The colours of the grid cells indicate their species richness. Saturated red colours indicate many species and dark blue colours indicate few species.

10.4   DATA ANALYSIS

The WORLDMAP software was kindly provided by Paul Williams (Natural History Museum, London) with a 0.25° grid projection for Borneo and adjacent smaller islands (Williams 2001). This grid resolution is thought to be fine enough to allow for conservation planning and it is probably the finest geographical resolution permissible for any tropical region (Fjeldså 2000). The rattan distribution data were imported into WORLDMAP while retaining distinction between specimen records of the taxa and their estimated distribution. In analyses using WORLDMAP, the default settings of the programme were used throughout, unless otherwise indicated.

10.4.1   DISTRIBUTION, ENDEMISM AND PROTECTION STATUS OF THE RATTAN TAXA

By examining their estimated distributions and with input on the distribution of rattan taxa outside Borneo, the Bornean rattan taxa were described with regard to their individual distribution and endemism patterns. (For terminology, see Endnote 2.) The protection status was assessed for all rattans with narrow distribution following the IUCN Red List 2001 Categories and Criteria (version 3.1) (http://www.redlist.org/info/categories_criteria. html). In practice, only a few criteria were applicable to the data set. The criteria used are 'critically endangered', 'endangered' or 'vulnerable'. The assessments were based on the *area of occupancy* of a given taxon and whether the habitat area is likely to decrease. A taxon area of occupancy was calculated as a summation of the extent of natural vegetation in the grid cell(s) in which the taxon is estimated to occur. In assessment of whether a taxon habitat was declining, we used the somewhat arbitrary assumption that taxa, which are only estimated to occur in grid cells that have 75% or less natural vegetation, were subject to habitat decline. If a species was found in one well-protected grid cell (at least 30% covered by reserves; see following), it was not considered subject to habitat decline.

10.4.2   RICHNESS AND ENDEMISM PATTERNS

Using the estimated distributions of the rattans, the WORLDMAP software was used to produce maps illustrating patterns of species richness and range-size rarity (endemism richness) on Borneo (Figure 10.5, Figure 10.6 and Figure 10.7).

Richness of a grid cell was defined simply as the number of taxa occurring in the cell. The range-size rarity score of a grid cell was calculated as the accumulated range-size rarity scores of all taxa estimated to occur in the cell (WORLDMAP setting: rarity by inverse range-size) (Williams 2001). A map was also produced illustrating taxon richness by only using specimen records. Regarding range-size rarity, the resulting map was 'smoothed' twice (WORLDMAP settings: neighborhood smoothing=mean) to produce a map giving more emphasis to general patterns of range-size rarity than to small-scale fluctuations between grid cells.

10.4.3   COLLECTION INTENSITY

In order to calculate the total number of rattan collections made in all grid cells on Borneo, all rattan specimens with a geographical reference were imported into WORLDMAP without identifications to produce a map showing the total number of rattan specimens collected

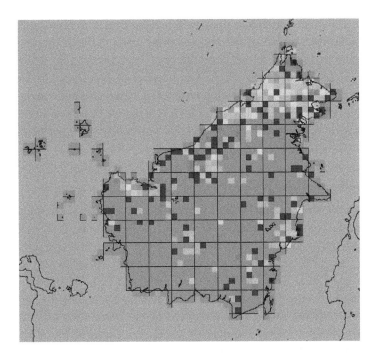

FIGURE 10.6 **(Colour Figure 10.6 follows p. 180.)** WORLDMAP illustration showing the richness of rattans on Borneo based on confirmed records only. The colours of the grid cells indicate their species richness. Saturated red colours indicate many species and dark blue colours indicate few species.

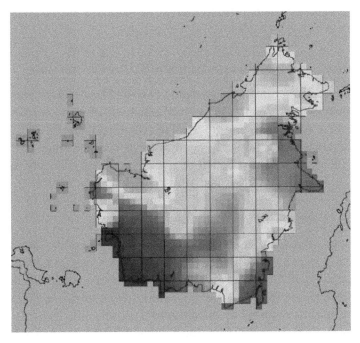

FIGURE 10.7 **(Colour Figure 10.7 follows p. 180.)** WORLDMAP map showing range-size rarity patterns of rattans on Borneo based on estimated distributions. Saturated red colours indicate high range-size rarity scores and the dark blue colours indicate low values.

in the grid cells. The information was displayed as a bar diagram. Pearson Product Moment Correlation between the number of rattan collections made in the grid cells and the number collected, the estimated rattan richness and the range-size rarity score was calculated using SigmaStat 1.0 (Jandel Corporation 1993). The number of specimens collected in the grid cells was plotted against the corresponding number of rattan taxa recorded. A curve was fitted using the regression function of Sigma Plot 8.0 using an 'exponential rise to maximum; single, two parameters' equation ($y = a(1 - e^{-bx})$, x = number of specimens collected, y = number of rattan taxa recorded) (SPSS 2002).

10.4.4 MAXIMIZING RATTAN PROTECTION

The rattan taxa were assumed to be well protected in grid cells with at least 30% reserve coverage (following Lund and Rahbek 2000) and were assumed effectively conserved if they were estimated to occur in at least one of these well-protected grid cells. The effectiveness of the current set of well-protected grid cells was then calculated as the proportion of rattan taxa occurring in at least one of the well-protected cells. A similar calculation was made that included also the proposed reserves in the calculation of the cell reserve coverage. The proportion of the rattan taxa protected in the current set of well-protected grid cells was compared to the average number of rattan taxa captured by randomly selecting a similarly sized set of grid cells (1000 simulations) and the number of rattan taxa captured by selecting a similarly sized set of grid cells by using the *near-maximum coverage* algorithm of WORLDMAP (Williams 2001; Endnote 1). The Monte Carlo test implemented in the random-selection tool in WORLDMAP (Williams 2001) was used to investigate whether there was a significant difference between the number of rattan taxa captured by the current set of well-protected grid cells and by the random-set simulation (P = 0.05, two-tailed test).

10.4.5 IDENTIFYING A MINIMUM SET OF AREAS FOR RATTAN CONSERVATION

The near-minimum set algorithm implemented in WORLDMAP (Williams 2001; see Endnote 2 for explanation) was used to calculate different sets of priority areas for rattan conservation on Borneo. Near-minimum set analyses were made for (1) all taxa; (2) endemic taxa; and (3) threatened endemic taxa, using estimated distributions. Separate analyses were made that did not allow for selection of grid cells that are currently well protected. A near-minimum set analysis was also made using only specimen records.

10.4.6 COMPARING AREA SELECTION METHODS

The efficiency of the near-minimum set algorithm was compared to (1) random selection of grid cells; (2) selection of areas of high taxon richness (area species richness); (3) selection of areas of high range-size rarity endemism richness by inverse range size; and (4) selecting areas using the greedy set algorithm, rejecting within-set redundant areas (Williams 2001). The number of grid cells required by the different area selection methods to represent all species at least once was compared. Also, the number of taxa captured by the different area selection methods was compared, when these were set only to select the amount of grid cells required by the near-minimum set algorithm to represent all taxa at least once. The Monte Carlo test implemented in the random-selection tool in WORLDMAP (Williams

2001) was used to investigate whether the differences between the numbers of rattan taxa captured by the various area selection methods and by the random-set selection were significant ($P = 0.05$, one-tailed test).

10.4.7   OPTIMIZING AREA SET BY VEGETATION DATA

To select the grid cells with the largest possible natural vegetation cover, data on the current level of land degradation (defined as the proportion of land degraded or cleared in the grid cells) were imported into WORLDMAP and converted to cost data such that high levels of land degradation in a grid cell resulted in high cost value. The near-minimum set algorithm was then set to optimize area selection by cost data (WORLDMAP setting: near-optimum area set: optimize area set by cost data). In practice, the algorithm would thereby choose the grid cell with the most natural vegetation when two grid cells were otherwise equally eligible for selection.

10.4.8   AVOIDING RELIANCE ON HIGHLY DEGRADED GRID CELLS

Grid cells with less than 10% natural vegetation were excluded to allow for selection of grid cells more feasible for conservation. This analysis could not therefore capture those taxa that were only represented in grid cells with less than 10% natural vegetation. However, this analysis would ensure that all other taxa would be represented in at least one grid cell with 10% or more natural vegetation.

10.4.9   THE FINAL SET OF PRIORITY AREAS FOR RATTAN CONSERVATION

The final set of priority areas for rattan conservation on Borneo was calculated in four steps:

1. A near-minimum set analysis using estimated distributions of all taxa, including all grid cells and optimizing vegetation cover in the area set, was conducted to select a set of grid cells that represented all taxa at least once.
2. A near-minimum set analysis similar to that in step 1 was conducted that only included grid cells with at least 10% natural vegetation.
3. The results of step 1 and step 2 were compared, and those grid cells that were selected in step 2 but not in step 1 were identified.
4. The grid cells selected in step 1 and those identified in step 3 were combined to form the final set of priority areas for rattan conservation on Borneo.

Following this procedure, the final set of grid cells represented all taxa at least once, where possible, in at least one grid cell with 10% or more natural vegetation.

10.5   RESULTS: RATTANS OF BORNEO AND THEIR PROTECTION STATUS

In total, 140 rattan species belonging to eight genera (Table 10.4) are recorded on Borneo, including the monotypic, endemic genus *Retispatha*. Most of the species belong to the genus *Calamus* (79 species). Including varieties, the total number of taxa in this study is 144, of which 98 (68.1%) are endemic to Borneo (Dransfield and Patel 2005). Most of the rattan taxa (86 taxa or 59.7%) are widespread or very widespread, whereas 21 (14.6%) have local distribution and 37 (25.7%) have narrow distribution (see Endnote 2 for definitions of terms used to describe taxic distributions). The great majority of the narrowly distributed taxa are endemic to Borneo.

Assessment of the protection status of the narrowly distributed taxa reveals that 27 (73.0%) of these are threatened (i.e., classified as critically endangered, endangered or vulnerable) following the IUCN Red List 2001 Categories and Criteria (version 3.1) (http:// www.redlist.org/info/categories_criteria.html). In total, this means that at least 18.8% of the rattans on Borneo and at least 23.5% of those endemic to the island are threatened to some degree. The 23 threatened endemic rattans are threatened globally because they are not found outside Borneo. (See Table 10.6 for a summary of the protection status assessment.) Three taxa are classified as critically endangered, which means that they are considered to be facing an extremely high risk of extinction in the wild. Two of the three taxa are endemic to southern Sarawak; the third occurs on Malawali Island immediately north of Borneo.

10.5.1 RICHNESS AND ENDEMISM PATTERNS OF RATTANS ON BORNEO

The taxon richness map (Williams 2001) based on estimated distributions of all rattan taxa on Borneo is shown in Figure 10.6. The most taxon-rich areas (red) are two grid cells in Brunei, both encompassing 67 rattan taxa or 48% of the total taxon richness of Borneo. The most taxon-poor areas are the coastal regions in south-western Borneo, where only 10 rattan species are estimated to grow. The overall richness pattern indicates that the northern and north-western regions contain more species than the southern and the eastern ones. A different richness map based on specimen data only is shown in Figure 10.6. This illustrates the enormous variation in collected rattans in different regions of Borneo.

The map indicates that larger endemism richness localities are found in the south-western Sarawak, Brunei and the central parts of Sabah. Smaller areas of high range-size rarity include the islands immediately north of Borneo (Banggi and Malawali); the area around Bukit Silam, Sabah; Banjarmasin, Pontianak and Balikpapan areas, Kalimantan. In Figure 10.7 the number of rattan specimens collected in the grid cells varies greatly. In 71.8% of the 1087 grid cells, no rattans have ever been collected; in only 5.7% of the grid cells more than 20 rattan specimens have been collected. A Pearson Product Moment Correlation test (Table 10.7) indicates a highly significant and strong positive correlation between the number of specimens collected in a grid cell and the number of rattan taxa recorded (Table 10.6).

A similar correlation result was obtained by excluding from the test the 71.8% grid cells in which no rattan collections had been made. There is also highly significant positive correlation between the number of specimens collected in a grid cell and its estimated rattan taxon richness and estimated range-size rarity score. However, the correlation with the latter two variables is weaker than with the former two. Correlation test results are summarized in Table 10.6. Figure 10.8 shows the number of rattan collections in a grid cell plotted against its corresponding number of rattan taxa, with a fitted curve.

10.5.2 AREA SELECTION MAXIMIZING RATTAN PROTECTION

The 70 nature reserves on Borneo cover 8% of the entire land surface of the island (MacKinnon, pers. comm.). A total of 449 (41%) of the 1087 grid cells included in this study contain smaller or larger parts of at least one reserve, and in 91 grid cells the reserve covers at least 30%.

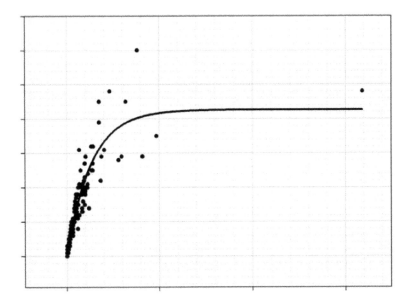

FIGURE 10.8   The number of rattan specimens in a grid cell plotted against the corresponding rattan taxa, with a fitted curve.

Assuming that all rattan taxa are sufficiently protected in areas with at least 30% coverage by reserves, these 91 grid cells are defined as representing the current set of well-protected grid cells on Borneo. In these 91 grid cells, 103 (71.5%) rattan taxa are protected.

Assuming that the proposed reserves on Borneo were all implemented, the number of well-protected grid cells would rise to 175. This enlarged set would protect 110 (76.4%) of the 144 rattan taxa found on the island.

If 91 grid cells (the current number of well-protected grid cells) are chosen randomly, only 108 (75.0%) rattan taxa are captured. The difference between the level of protection exercised by the current set of protected grid cells and the random-set simulation is insignificant. Hence, the current set of well-protected grid cells is no better than a random selection.

Using the near-maximum coverage algorithm (Williams 2001) to choose 91 grid cells, all taxa are protected. The rarest 37 taxa (25.7%) (i.e., all narrowly distributed) are protected in all the grid cells where they occur, and the remaining taxa are all represented in at least seven grid cells (Table 10.8). It is clear that this set of 91 protected grid cells is far more effective in protecting rattan taxa than the current set of 91 well-protected grid cells. The discrepancy in coverage between the two sets of grid cells is illustrated by an overlay in Figure 10.9. Only seven grid cells are in common between the two sets of grid cells.

10.5.3   A PRIORITY SET OF AREAS FOR RATTAN CONSERVATION

A near-minimum set analysis, based on estimated distributions and including all taxa and grid cells, required 23 grid cells to represent all taxa at least once (16 grid cells are irreplaceable, 3 are partly flexible and 4 are fully flexible). An irreplaceable grid cell is required for achieving the conservation goal. A partially flexible grid cell is required for achieving the conservation goal most efficiently in terms of the required number of grid cells. A flexible grid cell can be replaced by another without compromising the goal efficiency).

TABLE 10.6
Priority Areas for Rattan Conservation on Borneo

Name of grid cell	Locality of grid cell	Natural vegetation (in % of grid cell)	Reserve coverage (in % of grid cell)	Narrowly distributed taxa (extent occurrence in brackets)	Recorded/ estimated occurrence of taxa
Pulau Banggi	7'-7'15N; 117'-117'15E	17.65	0.00	D. banggiensis J. Dransf. (1)	6/48
Pulau Malawali	7'-7'15N; 117'15-117'30E	0.00	0.00[a]	C. malawaliensis J. Dransf. (1)	4/47
Kinabalu Park	6'-6'15N; 116'30-116'45E	71.43	70.10	P. elongata Mart. Ex. Bl. (4); C. mesilauensis J. Dransf. (3)	48/57
Kuala Labuk	6'-6'15N; 117'30-117'45E	19.35	0.00	C. sabensiss Becc. (1)	1/50
Tawai	5'30-5'45N; 117'15-117'30E	41.94	0.00	C. laevigatus var. serpentinus J. Dransf. (6); C. hepburnii J. Dransf. (1); D. serpentina J. Dransf. (6)	7/53
Northern Brunei	4'45-5'N; 115'-115'15E	70.00	16.94	C. temburongii J. Dransf. (2)	8/62
Bukit Silam	4'45-5'N; 118'-118'15E	55.32	0.00	C. microsphaerion Becc. Vel Valde (2); C. diepenhorstii var. major J. Dransf. (3); D. serpentina J. Dransf. (6)	12/52
Western Brunei	4'15-4'30N; 114'15-114'30E	80.65	7.58	C. axillaris Becc. (5); C. maiadum J. Dransf. (4)	20/67
Mulu National Park	4'-4'15N; 114'45-115E	100.00	66.84	C. nielsenii J. Dransf. (1)	48/69
Sungai Bahau	2'30-2'45N; 116'15-116'30E	100.00	15.13	C. tomentosus Becc. (1)	7/35
Samnsan	1'45-2'N; 109'30-109'45E	56.25	10.66	C. poensis Becc. (2); C. pygmaeus Becc. (7); D. draco (Willd.) Bl. (1)	21/47

Kubah National Park	1'30–1'45N; 110'–110'15E	12.00	7.93	*C. pygmaeus* Becc. (7); *C. conjugatus* Furtado (2); *C. crassifolius* J. Dransf. (2)	60/68
Apo Kayan	1'30–1'45N; 115'–115'15E	75.81	0.00	*D. pumila* Volkenburg (1)	20/47
Semengoh	1'15–1'30N; 110'15–110'30E	0.00[a]	1.04	*C. hypertricosis* Becc. (1)	32/64
Tebebu Area	1'–1'15N; 110'15–110'30E	0.00[a]	0.00	*D. unijuga* J. Dransf. (1)	18/45
Samaharan East	1'–1'15N; 110'45–111'E	14.52	0.00	*C. paulii* J. Dransf. (1); *C. sabalensis* J. Dransf. (2); *P. moorei* J. Dransf. (1)	39/62
Sri Aman West	1'–1'15N; 111'–111'15E	15.87	0.00	*C. psilocladus* J. Dransf. (1); *C. sabalensis* J. Dransf. (2)	6/60
Semitau Area	0'30–0'45N; 111'45–112'E	50.00	0.00	*C. tappa* Becc. (5); *C. impar* Becc. (1)	23/38
Bukit Batu Tenobang	0'15–0'30N; 113'–113'15E	100.00	0.00	*D. halleriana* Becc. (2)	10/43
Pontianak	0'–0'15S; 109'15–109'30E	26.98	0.00	*K. paucijuga* Becc. (3); *P. pygmaea* Madulid (1)	18/25
NW Balikpapan	0'45–1'S; 116'30–116'45E	80.65	34.59	*C. spectatissimus* Furtado (4)	2/40
Balikpapan	1'–1'15S; 116'45–117'E	26.79	34.59	*C. nigricans* Valkenburg (1); *C. fimbriatus* Valkenburg (1)	23/38
Banjarmasin	3'15–3'30S; 114'30–114'45E	9.84[a]	0.00	*C. schistoacanthus* Bl. (2); *C. tapa* Becc. (5); *C. winklerianus* Becc. (1)	25/36
Danau Sentarum[b]	0'45–1'N; 112'30–112'15E	23.81	0.00	*C. schistoacanthus* Bl. (2); *C. tapa* Becc.	5/43
Gunung Menyapa Area[b]	0'45–1'N; 116'30–116'15E	100.00	0.00	None	0/41
SW Palangkaraya[b]	2'45–3'S; 113'30–113'45E	96.83	0.00	None	0/18

[a] Areas excluded for the second near-minimum set analysis because they have less than 10% natural vegetation (see Section 4.4 for additional details).

[b] Additional areas selected in the second near-minimum set analysis (see Section 4.4 for additional details).

TABLE 10.7

Changes in Natural Vegetation on Borneo

Current vegetation (sq. km)	Brunei Area	%	Sabah Area	%	Sarawak Area	%	Kalimantan Area	%	Borneo Area	%
Cleared	1188.5	20.24	34739.1	46.29	57321.7	45.69	222078.8	40.78	315328.1	41.99
Degraded	787.3	13.41	6024.5	8.03	0.0	0.00	0.0	0.00	6811.9	0.91
Freshwater swamp	77.0	1.31	832.0	1.11	187.0	0.15	18283.1	3.36	19379.2	2.58
Heath forest	0.0	0.00	94.7	0.13	652.2	0.52	37584.7	6.90	38331.7	5.10
Limestone	0.0	0.00	23.4	0.03	75.9	0.06	1710.3	0.31	1809.6	0.24
Mangrove	202.6	3.45	699.7	0.93	972.3	0.78	5882.6	1.08	7757.2	1.03
Peat swamp	1262.9	21.50	38.6	0.05	5717.3	4.56	22204.1	4.08	29222.9	3.89
Tropical montane evergreen	74.6	1.27	3071.4	4.09	14762.6	11.77	27951.3	5.13	45859.9	6.11
Tropical wet evergreen	2280.4	38.83	29408.5	39.19	45762.5	36.48	208915.5	38.36	286366.9	38.13
Not classified	0.0	0.00	118.3	0.16	0.0	0.00	14.4	0.00	132.7	0.02
Total area	**5873.2**	**100.00**	**75050.3**	**100.00**	**125451.6**	**100.00**	**544624.9**	**100.00**	**751000.0**	**100.00**

Original vegetation (sq. km)	Brunei Area	%	Sabah Area	%	Sarawak Area	%	Kalimantan Area	%	Borneo Area	%
Freshwater swamp	93.0	1.56	2601.3	3.47	299.5	0.24	38345.0	7.04	41338.7	5.50
Heath forest	88.0	1.48	278.7	0.37	1381.7	1.10	78454.4	14.40	80202.8	10.68
Limestone	0.0	0.00	106.6	0.14	384.9	0.31	2075.5	0.38	2567.1	0.34
Mangrove	252.4	4.24	4712.8	6.28	3186.3	2.54	19515.1	3.58	27666.6	3.68
Peat swamp	1675.5	28.17	314.0	0.42	13620.5	10.86	43535.5	7.99	59145.5	7.87
Tropical montane evergreen	74.6	1.25	3500.7	4.66	16899.5	13.47	28848.2	5.30	49323.0	6.57
Tropical wet evergreen	3765.1	63.29	63539.5	84.66	89663.0	71.48	333884.9	61.30	490852.4	65.35
Not classified	0.0	0.00	0.1	0.00	10.8	0.01	0.0	0.00	10.9	0.00
Total area	**5948.5**	**100.00**	**75053.7**	**100.00**	**125446.2**	**100.00**	**544658.7**	**100.00**	**751107.1**	**100.00**

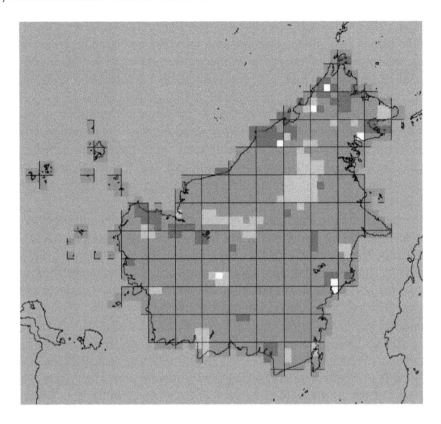

FIGURE 10.9 (Colour Figure 10.9 follows p. 180.) WORLDMAP between 91 grid cells in which 30% is covered by protected areas (green colour) and grid cells selected in a near-maximum coverage for 91 grid cells (blue colour). White cells indicate overlap.

TABLE 10.8
Results of Four Different Product Moment Correlation Tests

Pearson product moment correlation test	Correlation coefficient	P-value	No. of samples
Rattan specimens collected vs. rattan taxa collected	0.703	5.27E-163	1087
Rattan specimens collected vs. rattan taxa collected (excluding grid cells without rattan collections)	0.681	3.35E-043	307
Rattan specimens collected vs. estimated rattan taxon richness	0.254	1.82E-17	1087
Rattan specimens collected vs. estimated range-size rarity score	0.415	1.65E-46	1087

Note: Pairs of variables with positive correlation coefficients and p-values below 0.05 tend to increase together.

The differences between this algorithm and other area selection algorithms implemented in WORLDMAP were investigated. A random selection of 23 areas (1000 replicates) captured on average 62% of the rattan taxa. The Monte Carlo test revealed that any selection of 23 grid cells that captures more than 69% of the taxa is significantly better than a random selection. Selecting 23 grid cells based on their number of rattan taxa capture 86% of the rattan taxa, and 214 grid cells were required to represent all taxa at least once (all cells were located in the Brunei area and in south-western Sarawak). Selecting 23 grid cells based on their rattan range-size rarity score (rarity by inverse range size), 99% of the rattan taxa were captured, and 33 areas were required to represent all taxa at least once. Using a greedy set algorithm, 23 grid cells were required to represent all taxa at least once.

In choosing a minimum set of priority areas for rattan conservation on Borneo, it is clear that the near-minimum set algorithm is more efficient than random selection and selection by taxon richness or range-size rarity score. In this study, the near-minimum set algorithm and the greedy set algorithm are equally efficient. It is worth noting that the sets of grid cells selected by the greedy set algorithm and near-minimum set algorithm (optimizing area set for natural vegetation cover) are exactly the same.

With the near-minimum algorithm set to optimize areas for natural vegetation cover in the grid cells — thereby favouring grid cells with high natural vegetation cover — analysis resulted in a set of 23 grid cells (16 grid cells are irreplaceable and 7 are partly flexible — that is, they could be different cells). This set of grid cells differed marginally from the near-minimum set analysis without optimization for natural vegetation cover because all but three selected grid cells were identical in the two analyses.

The consequences of analysing subsets of the data with the near-minimum set algorithm (optimizing area set for natural vegetation cover) were also investigated. Analysis based on the 98 rattan taxa endemic to Borneo selected 20 grid cells to represent all endemics at least once. All of these grid cells were also selected in the near-minimum set analysis of all taxa. Analysis based on the 27 narrowly distributed and threatened rattan taxa selected 18 grid cells to represent all narrowly distributed and threatened rattan taxa at least once. All of these grid cells were also selected in the near-minimum set analysis of all taxa. An analysis for all taxa, which preselected the 91 well-protected grid cells, selected 19 additional grid cells to represent all taxa at least once. All the 19 additional grid cells were also selected in the near-minimum set analysis without preselected grid cells. Using only specimen records (in contrast to estimated distributions in the preceding tests), a near-minimum set analysis of all taxa selected 26 grid cells to represent all taxa at least once. Out of these, 21 grid cells were also selected in the near-minimum set analysis using estimated distributions.

The subsequent selection of only one of these different sets of grid cells as a preliminary set of priority areas for rattan conservation on Borneo was based on the following assessments:

1. The actual level of protection of rattan taxa inside reserves on Borneo is not sufficiently well known.
2. The level of protection of non-endemic rattan taxa outside Borneo is not sufficiently known.
3. The conservation goal is to protect all rattan taxa and not only those that are currently classified as threatened.
4. The estimated distributions of the rattan taxa of Borneo developed in this study are a better reflection of the actual distributions than the confirmed specimen records of rattans alone.

Thus, the set of 23 grid cells resulting from the near-minimum set analysis of all taxa using their estimated distributions, optimizing area set for natural vegetation cover and without preselecting grid cells was selected as a preliminary set of priority areas for rattan conservation on Borneo.

Four of the chosen grid cells had less than 10% of their natural vegetation left (Mac-Kinnon, pers. comm.). A near-minimum set analysis, which excluded all grid cells with less than 10% natural vegetation cover, selected 22 grid cells to represent all taxa found outside the excluded grid cells at least once. Since three of these grid cells were not found in the preliminary set of preceding priority areas, they were added to the final set of priority areas for rattan conservation on Borneo.

In the final set of 26 grid cells, all rattan taxa are represented at least once. All but four rattan taxa are represented at least once in a grid cell with 10% or more natural vegetation cover; the exceptions are *Calamus malawaliensis*, *Calamus hypertrichosus*, *Daemonorops unijuga* and *Calamus winklerianus*. These four taxa all have unique records for grid cells with less than 10% natural vegetation. The final list of priority areas for rattan conservation on Borneo is shown in Table 10.7 and Figure 10.10.

10.6 DISCUSSION

10.6.1 DIVERSITY AND ENDEMISM

Of the world's approximately 580 rattan species, 140 species (24%) are found on Borneo. Since 95 of the rattan species are endemic to the island, 16% of all rattan species are only found on Borneo. Given the relatively wide distribution of rattans worldwide, these numbers may be seen as indicating high levels of species richness and endemism on Borneo.

In this study, 5045 rattan records were gathered from eight herbaria and through field-work. However, it is important to note that this data set most likely includes not all rattan specimens collected on Borneo. Within the time frame of this study, it was impossible to obtain rattan data from the Sarawak herbarium (SAR). We estimate that approximately 80% of all rattan collections at SAR are duplicated at Kew (K). Some rattan records from the Leiden herbarium (L) have not been included in the present study either. It is likely that the missing records result in underestimation of the distribution area of some taxa, particularly in Sarawak and Kalimantan. However, due to the large proportion of the existing rattan specimen data that are included and the diversity of sources they are gathered from, it is considered unlikely that adding the missing records would significantly change the results presented here.

10.6.2 DISTRIBUTION PATTERNS

There are great differences in the estimated taxon richness between the richest regions in Sabah, Brunei and Sarawak and the poorest regions in southern and especially south-western Kalimantan (Figure 10.9). The pattern in taxon richness is very much in agreement with the findings of Soepadmo and Wong (1995) on the tree flora of Borneo. The pattern of rattan taxon range-size rarity is fairly similar to that of taxon richness. It differs mainly in that south-western Sarawak and eastern Kalimantan appear to be distinct range-size rarity hotspots, whereas they are not distinct richness hotspots. Also, a number of smaller areas of high range-size rarity appear scattered in Kalimantan (Figure 10.10).

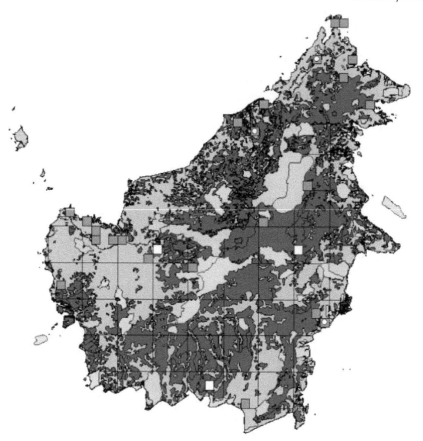

FIGURE 10.10 **(Colour Figure 10.10 follows p. 180.)** A map showing the locality of the 26 priority areas for rattan conservation on Borneo. Red quadrates are grid cells selected in the preliminary set of priority areas for rattan conservation. Yellow quadrates are additional grid cells selected in the *near-minimum set* analysis in which grid cells with less than 10% natural vegetation were excluded. Together, red and yellow quadrates make up the final set of priority areas for rattan conservation on Borneo. Selected grid cells with a white dot are well protected. Orange polygons are reserves and bright blue polygons are proposed reserves. Dark green illustrates natural vegetation and bright green illustrates cleared or degraded land.

It may be noted that, to some extent, the patterns of high richness and especially of rarity may be explained by easy access to the areas due to relatively good infrastructure (e.g. central Sabah), proximity to larger cities (south-western Sarawak and most of the hotspots of range-size rarity in Kalimantan) or extensive field collections (Brunei).

The Pearson Product Moment Correlation tests demonstrate a strong positive correlation between the number of rattan specimens collected in a grid cell and the grid cell's estimated taxon richness and range-size rarity score (Table 10.8). If one uses the number of rattan specimens collected in a grid cell as a surrogate for collection intensity, then the Pearson Product Moment Correlation tests support the earlier statement that diversity of taxon richness and range-size rarity are more correctly high points of collection intensity.

However, the number of collected rattan specimens is a far from a perfect surrogate and collection intensity would be measured more correctly as the number of man hours spent collecting rattans in a given area. In most parts of Borneo, there have been a very

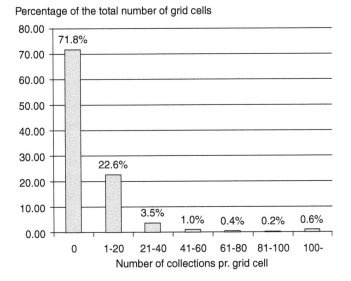

FIGURE 10.11   The number of collections made in the grid cells illustrated as intervals in the amount of collections (x axis) against the corresponding percentage of grid cells (y axis).

limited number of collection trips. Most botanists would collect one specimen of each taxon encountered, especially of taxa that are difficult or cumbersome to press. In poorly collected areas, most rattan taxa will therefore have been collected only once. Thus, there will be a very strong positive correlation between the number of collected specimens and the number of rattan taxa recorded. In the relatively few grid cells on Borneo that have been visited frequently by many different collectors, more duplication of efforts occur and the correlation is weakened because most taxa have been collected many times. This collection dynamic is illustrated in Figure 10.11. In areas where few collections have been made, there is a neat linear relationship between the number of collected specimens and the number of collected taxa, whereas the relationship is less obvious in grid cells where many collections have been made.

As illustrated in Figure 10.11, it is clear that many grid cells are poorly collected. In most cells, no rattans have been collected, and it is not likely that this reflects rattan distribution on Borneo. The scattered collection effort on Borneo is a limitation of the data set to consider when using it for conservation priorities. The collection bias, in part, may be compensated by using estimated distributions rather than specimen records for analysing the patterns of rattan diversity and distribution, since gaps in collection efforts are hereby effectively bridged. The danger in this regard is that the estimates are subjective and imprecise and the analyses become flawed. In this study, specimen records, one expert judgement and a set of guidelines were used to produce distribution maps. This largely ensures consistency in the estimated distributions. Following the guidelines, the distribution maps are relatively conservative estimates of the rattan distributions. The risk of overestimating rattan distributions is therefore considered to be limited.

A different method for estimating distribution maps would have been the use of computer modelling. To reliably estimate potential plant distributions through modelling, it is important to have very precise geographical references for the specimen records (Borchsenius and Skov 1999). As outlined in Endnote 1, a suite of tools and methods was necessary to geographically reference the specimens with a precision of plus or minus one grid cell. This is to indicate that the present specimen data set does not have very precise geographical reference, and thus estimating rattan distributions on Borneo using computer modelling would probably not have added much value.

10.6.3 Threatened Rattan Taxa

Despite the fact that 8% of Borneo is covered by reserves, at least 18.8% of the rattan taxa are threatened. The 23 endemic taxa that were assessed to be threatened in this study may all be added to the IUCN Global Red List (www.redlist.org). In comparison to this study, Pearce (1989) found 44 rattan taxa threatened in Sarawak. There can be several reasons for the difference between Pearce's study and the present one. Most importantly, only the narrowly distributed taxa are evaluated here. It is unlikely that any additional rattan taxa would presently qualify as critically endangered, but several may be assessed as endangered or vulnerable. This is to indicate that the proportion of threatened rattan taxa on Borneo may be somewhat higher than the number estimated here. Second, taxa threatened in Sarawak may not be threatened on Borneo as a whole. Some taxa threatened in Sarawak may not be threatened in other regions on the island and will not appear as threatened in this study. Third, Pearce used an older set of assessment criteria different from those applied here. Finally, it cannot be ruled out that the conservation assessment by Pearce has relieved the pressure on some rattan taxa in Sarawak. Given the low percentage of Sarawak covered by protected area, however, we find this an unlikely option.

The three taxa that are classified as being critically endangered are all found in areas characterized by very poor habitat and very poor protection; the most recent collections of them are from the late 1970s or the beginning of the 1980s. This may indicate that the taxa are extinct, but field studies are required in order to confirm this before the taxa may be classified as such (http://www.redlist.org/info/categories_criteria2001.html#categories).

In the interpretation of the assessment criteria in this study, a few assumptions were made that may influence the assessments. First, the area of occupancy of a taxon was defined as equal to the extent of natural vegetation in the grid cells where it occurs. This may be an overestimation because rattans need not occur in all suitable habitats in the cells. Hence, some species that are in fact threatened may not be evaluated as such.

The other important interpretation that has been made is that a taxon should experience continued habitat decline to be threatened. Bearing in mind the high rates of deforestation on Malaysia and Kalimantan (FAO 2001) and the future scenarios developed by GLOBIO (http://www.globio.info), it was assumed that areas with less than 75% natural vegetation are subject to human pressures likely to continue and thus cause continued habitat decline. This is a relatively crude assumption since the vegetation cover of a particular grid cell does not predict the dynamics of decline or increase in rattan habitat in the cell. The IUCN Red List criteria do allow for inferential assessments of habitat decline as an option (http://www. redlist.org/info/categories_criteria.html), but interpretation of 'continued habitat decline' as applied here may nevertheless in some cases lead to assessing some taxa as threatened that are not. The IUCN Red List criteria also stress, however, that uncertainties should not be used as an excuse for not assessing a particular taxa as threatened. Following the precautionary principle implemented in the IUCN Redlist Criteria, the uncertainty should favour the protection of the taxon (http://www.redlist.org/info/categories_criteria.html).

10.6.4 EFFECTIVENESS OF THE EXISTING RESERVE NETWORK

The network of reserves on Borneo is estimated to protect 71.5% of the island's rattan taxa. The reserves on Borneo differ greatly in their extent and legal status and probably also in how well they protect biodiversity. In this study, it was not feasible to examine which of the individual reserves is sufficiently well managed to maintain biodiversity. Furthermore, since high-quality lists of rattan taxa found in reserves only exist for a few of the reserves on Borneo (e.g., Dransfield 1984b; Pearce 1994; Gait et al. 1998; Boje 2000; Andersen et al. in press), the actual occurrence of rattans inside reserves could not be measured precisely. Therefore, it was assumed that rattan taxa are effectively protected if they occur in at least one grid cell that is covered by at least 30% reserves. This definition of well-protected grid cells is also used by Lund and Rahbek (2000), whereas Williams et al. (1996) define well-protected grid cells as cells with at least 50% coverage of reserves. Fjeldså and Rahbek (1997) define a well-protected grid cell as a cell with:

- at least three nominally different reserves;
- formal protection of at least one third of the area of the grid cell; and
- effective protection of the biologically most unique parts of the grid cell.

It may be hazardous to define grid cells as well protected based only on the number of reserves without considering the actual area coverage, since numerous reserves may not effectively protect biodiversity if each is very small.

The assumption that rattans are effectively protected in grid cells with at least 30% reserve coverage has limitations: rattans may not be well protected inside reserves due to illegal logging or exploitation or they may be well protected outside reserves. Nevertheless, it may be argued that the assumption is a useful tool in assessing the level of rattan protection on Borneo because: (1) grid cells with large reserve coverage may generally provide better protection of rattans than grid cells with small reserve coverage; and (2) rattan taxa that are estimated to occur in areas with at least 30% reserve coverage are likely to occur in the reserves (Lund and Rahbek 2000).

When the efficiency of the current reserve network is compared to alternative sets of grid cells, it becomes evident that the current selection of reserves is inefficient in protecting rattan diversity. The current set of well-protected grid cells is no better than a random selection and it fails to protect 28.5% of the taxa. Conversely, the near-maximum coverage algorithm is able to select a set of grid cells of the same size as the current set of well-protected cells that ensure representation of all taxa in at least seven grid cells, where possible.

The inefficiency of the current set of reserves is probably caused by the criteria for their selection. A large proportion of the area covered by reserves on Borneo is found in the highlands of the central parts of Borneo. These areas may have been nominated as reserves because they represent the land that nobody wants because these areas are far away from human conglomerations (Forest Watch Indonesia and Global Forest Watch 2002) and possibly not immediately well suited for agricultural purposes.

The great number of areas that are proposed reserves on Borneo would lead to a marginal increase in rattan protection on Borneo, if they were all implemented, despite the fact that the total reserve area would be doubled. The rationale behind creating these reserves is not known to the authors, but it seems likely that the proposed reserves are laid out in the same pattern as the existing reserves. They are largely located in remote areas of the highlands of central Borneo.

10.7 INCREASING RATTAN PROTECTION ON BORNEO: METHODS FOR SELECTING PRIORITY AREAS FOR RATTAN CONSERVATION

The two complementarity-based algorithms investigated in this study (near-minimum set and greedy set algorithms) proved to be most efficient in representing all taxa at least once (23 grid cells required) compared to random selection of grid cells, cells of high taxon richness (richness hotspots) or cells of high endemism (endemism hotspots, expressed as range-size rarity scores). Much evidence is provided in the literature that area selection methods based on complementarity are superior to hotspot-based methods.

The reasons for the inefficiency of the hotspot-based methods in the present study may be explained by the rattan distribution patterns on Borneo. The most taxon-rich areas are Brunei and the surrounding areas plus Kuching and surrounding areas. The 23 grid cells of high taxon richness are concentrated in two clusters of grid cells in these two areas. Despite the high taxon richness in these areas, this method fails to capture those taxa that are confined to other regions. Selecting 23 grid cells of high endemism captures all but one taxon. The selected grid cells are more evenly scattered over the island, and choosing grid cells of high endemism is better than choosing grid cells of high taxon richness. However, the limitations of this method are clearly illustrated by the fact that it requires selection of 10 extra grid cells to capture the last taxon.

The two complementarity-based algorithms (near-minimum set and greedy set) were equally efficient in representing all taxa at least once and they selected the same grid cells. The near-minimum set algorithm was chosen for area selection because it allows for optimizing area selection based on the current levels of land degradation in the grid cells, and the algorithm can also provide information on irreplaceability and flexibility of the cells.

In analysing different subsets of the data set, the near-minimum set algorithm demonstrated a high degree of congruence and robustness; most differences were in the inclusiveness of the sets of selected grid cells. This may at least partly be explained by a large number of grid cells (16 out of 23) that are irreplaceable. In this context, an irreplaceable grid cell is a cell that must be selected to achieve the conservation goal of representing all taxa at least once, due to the unique rattan taxon record(s) made in the cell. The near-minimum set algorithm will inevitably select irreplaceable grid cells when the unique taxa they harbour are included in the analysis.

10.7.1 PROPOSED SET OF PRIORITY AREAS FOR RATTAN CONSERVATION

The proposed set of priority areas for rattan conservation on Borneo consists of 26 grid cells, which are required to represent all rattan taxa at least once, where possible, in at least one grid cell with 10% or more natural vegetation. Only three of these priority grid cells are currently well protected (i.e., covered by at least 30% reserves). The priority areas are scattered throughout all regions on Borneo with a majority in the northern and north-western regions. As mentioned earlier, 16 of the selected grid cells are irreplaceable, whereas the latter 10 are partially flexible. A partially flexible gird cell is required to achieve the conservation goal more effectively (Williams 2001). In this context, this means that excluding partially flexible grid cells would result in a requirement for a larger number of grid cells or selection of grid cells with poor vegetation. No grid cells are fully flexible and thus there are no alternative sets of grid cells that are equally efficient and appropriate, measured in terms of the number of selected cells and the proportion of natural vegetation in the cells.

Up to 88% of the rattan taxa estimated to occur in the selected grid cells have actually been collected in them; the average is 33%. This may be seen as a measure of certainty with regard to the actual rattan flora in the selected grid cells. Regarding age of the information in the data set, eight out of ten specimen records are made after 1970, about half after 1990. These figures apply to the total data set but are probably also valid for the 26 priority areas. The natural vegetation cover in the priority grid cells varies from 0 to 100%; the average is 48%.

All these gross figures, combined with the preceding considerations, reveal that this set of areas represents all rattan taxa at least once and may be the most appropriate set of priority areas for rattans on Borneo. There are, however, uncertainties as to whether conservation of these grid cells will effectively conserve all rattan taxa. Some taxa may have gone extinct and some selected areas may not harbour all the rattan species they are supposed to protect.

More flexibility might be added to the data set by selecting a set of priority areas that represent all taxa more than once, where possible. This would result in a somewhat larger set of priority areas. Representing all taxa more than once would probably result in more firm protection of many taxa with narrow or local distributions, but for those taxa that occur only in a single grid cell, this approach does not help; in fact, selecting more areas may divert focus from the highest priority areas.

10.7.2   REGIONAL OUTLOOK ON RATTAN CONSERVATION: BORNEO'S CONTRIBUTION TO RATTAN CONSERVATION IN SOUTHEAST ASIA

The conservation of the endemic rattan taxa on Borneo is primarily the responsibility of the countries that share Borneo. Only efforts on Borneo can ensure that these taxa do not go extinct in the wild. If all rattan taxa on Borneo were to become effectively protected, 24% of all rattan species in the world would be conserved. It is clear from this that conserving the rattans on Borneo would significantly contribute to worldwide rattan conservation. A total of 46 of the rattans on Borneo are also found outside the island, particularly elsewhere in Southeast Asia. Conserving these shared taxa may effectively complement the rattan conservation efforts elsewhere in the region.

10.7.3   FUTURE RATTAN RESEARCH IN SOUTHEAST ASIA

The methods developed and applied in this study for identifying priority areas for rattan conservation may well be duplicated in other parts of Southeast Asia. Since priority setting is useful not only in conservation but also in research, the question is which areas in Southeast Asia are most likely to be particularly important for conservation of rattans.

Williams and Gaston (1994) find that higher taxon richness is a good predictor of wholesale species richness in several groups of organisms and areas. There is good indication that the genus richness on Borneo is a good predictor of the lower taxon level richness. Assuming that this relationship is also true for the rest of Southeast Asia, data at the genus level may be used to set priorities for future rattan research. The two genera not found on Borneo are found in Peninsular Malaysia. This indicates that identifying priority areas for rattan conservation in the Malay Peninsula (Peninsular Malaysia, Singapore and southern Thailand) is an appropriate next step in conserving the rattans in Southeast Asia. A manual for rattans exists for the Malay Peninsula (Dransfield 1979, 1981) that indicates sufficient data are available on rattan distributions in this area to allow application of complementarity methods in selecting priority areas.

10.8   FROM ANALYSIS TO ACTION: PRIORITIES AND FEASIBILITY

The proposed set of priority areas should not be seen as a manual for rattan conservation on Borneo. Identifying priority areas as undertaken here is only the first step in planning practical rattan conservation. The feasibility of taking appropriate conservation measures must be investigated for each priority area individually through field studies. The application of *triage*, as described before, would probably assist well in screening the priority areas and to set priorities within the selected set of areas. Those grid cells that are well protected or have a very large natural vegetation cover may not require urgent measures. On the contrary, some of the priority areas apparently have very little or no natural vegetation; in these cells, field studies should be conducted to document their current rattan flora and the actual state of the vegetation before conservation measures are taken. Such field studies could lead to identification of very specific areas for conservation or possibly to the conclusion that effective conservation action is not feasible. Field studies may also be undertaken to update our knowledge about threatened taxa, particularly those that are critically endangered. This would be an important first step in estimating the feasibility of securing these taxa.

The actual cost of taking appropriate conservation measures may vary greatly between different areas. In this study, data were included on land degradation in the analysis to select those grid cells with most natural vegetation. This approach can be seen as an indirect way of including the cost of conserving the rattans in the grid cells because conservation measures may be less expensive in areas where the vegetation is relatively untouched. However, detailed knowledge on the real cost of managing (and possibly acquiring) land for conservation purposes is crucial. Calculating the cost for each of the 26 priority areas will be an important and necessary next step in moving towards concrete conservation measures. If some areas are excessively expensive to conserve, alternatives may be sought.

10.8.1   How to Protect Rattan Diversity in Priority Areas

As mentioned earlier, it is beyond the scope of this study to address in detail the important issue of how to manage biodiversity in priority areas. The present study may nevertheless provide valuable guidance. Many of the priority areas for rattan conservation are relatively highly populated or developed into agriculture. Thus, the classical concept of turning an area into a national park or giving it similar reserve status may not be appropriate or feasible in many of the priority grid cells. Instead, biodiversity management authorities may focus on implementing a range of measures aimed at achieving sustainable use of natural resources in the priority areas. In this regard, it seems important to address the harvesting of rattans in protected and unprotected forests. Legal or illegal exploitation of rattans for local or commercial purposes beyond sustainable levels poses a serious threat to rattans, even in areas with legal protection (Kiew and Pearce 1990).

10.8.2   COLLABORATIVE EFFORTS OF THE STATES ON BORNEO

Conservation action should be taken at the right level. Borneo forms a logical geographical unit. In this study, conservation priorities are therefore set for Borneo as a whole. The four different political units on Borneo — Brunei, Kalimantan, Sabah and Sarawak — may choose to use the overall priorities set for Borneo to take conservation actions individually and thereby contribute to the overall conservation of rattans. The conservation of the taxa that are endemic to one of the political units may be seen as their prime responsibility, whereas conservation of taxa shared with more than one political area requires collaboration. Information on the conservation status of such shared taxa should be exchanged and conservation efforts should be coordinated to ensure that no taxa are neglected.

In some cases, it may even be desirable to establish cross-border reserves to ensure large enough reserves to allow for long-term conservation of certain taxa. Cross-border collaboration is already taking place on Borneo between Indonesia and Malaysia through several projects funded by the International Tropical Timber Organization in the adjoining Lanjak Entimau Wildlife Sanctuary (Sarawak) and Betung–Kerihun National Park (Kalimantan) (http://www.itto.or.jp/inside/report2001/annex3.html). Such collaboration may be repeated elsewhere on Borneo based on common conservation priorities.

10.8.3   IMPLEMENTING INTERNATIONAL BIODIVERSITY AGREEMENTS ON BORNEO

Effective conservation of rattans on Borneo would be a direct contribution to the overall implementation of the Convention on Biological Diversity (CBD) and its provisions regarding *in situ* conservation and sustainable use of biological diversity (Secretariat to the Convention on Biological Diversity 2001b). The present study can also be seen as a contribution to the implementation of the recently adopted Global Strategy for Plant Conservation on Borneo (Convention on Biological Diversity 2002a), since it documents plant diversity and identifies gaps in plant conservation on Borneo.

The recent work programme on forest biological diversity of the CBD expresses a need for urgent action for forests that are threatened, important for biodiversity, and have potential for conservation and sustainable use (Convention on Biological Diversity 2002b). The Plan on Implementation of the World Summit on Sustainable Development suggests promoting and supporting initiatives for areas that are essential for biodiversity (http://www.johannesburgsummit.org). The present study may be seen as a pilot study to identify such areas. Evidently, more studies are required to identify areas that are important to a larger part of biodiversity, but similar methods may be used.

10.8.4   FEEDBACK TO INTERNATIONAL BIODIVERSITY-RELATED PROCESSES

Several methods of identifying the most important areas for rattan conservation have been tested in this study. Methods based on the principle of complementarity were found to be superior to other approaches — for example, hotspot-based methods. This is not new because many other studies within the last decade have reached the same conclusion. Despite this, some of the recent international agreements explicitly mention 'species richness' as a criterion for selection of priority areas and state that hotspots are areas that should receive particular attention.

There seems to be a need for communicating sound scientific findings regarding area selection methods to relevant international biodiversity-related policy development processes, most importantly the Convention on Biological Diversity. The message could be that, whenever possible, complementarity-based area selection methods should be applied in conservation planning to ensure effective and efficient identification of areas of particularly importance. Getting such a message across to the international biodiversity policy community would help ensure that the advice given to governments and other stakeholders involved in biodiversity conservation would become as targeted and precise as possible.

10.8.5 Towards a Set of Priority Areas for Biodiversity Conservation on Borneo

It is unlikely that any biodiversity conservation authorities base their selection of priority areas solely on rattans, nor would it be advisable. Using one group of organisms as indicators for wholesale biodiversity conservation should be done with great caution. On the other hand, it is extremely difficult to investigate the diversity and distribution patterns of all the groups of organisms present on Borneo. Most groups are very poorly studied and to insist on gathering data on all or most groups of organisms prior to selecting priority areas would effectively mean not to set any priorities at all. The best and most pragmatic surrogate that may be used for selecting priority areas for wholesale biodiversity conservation may be to use a limited number of different groups of organisms that are well studied and have well-documented distributions.

Recently, Stabell (2002) studied the diversity and distribution patterns of birds on Borneo using WORLDMAP, and investigations on butterflies and frogs are currently ongoing at the Institute for Tropical Biodiversity and Conservation, Universiti Malaysia Sabah (Dawood 1999; Effendi, pers. comm.; Boon Hee, pers. comm.). At the time of completion of this study, it was possible to make a preliminary comparison among rattans, birds and butterflies with regard to congruence in richness patterns and priority areas.

The bird data set includes distributions for 414 taxa (Stabell 2002). Figure 10.12 is an overlay of the richness maps of birds and rattans and illustrates high congruence in rattans and birds in the central parts of Borneo, especially in areas near the coast in Sarawak, Brunei and Sabah. The 26 grid cells selected as a priority set of areas for rattan conservation capture 385 (93.0%) of the bird taxa. This may indicate that the priority areas selected for rattans to some extent are also priority areas for birds. It is important to note, however, that inaccuracies in several of the bird distribution maps have been identified (Fjeldså, pers. comm.). Also, the grid system used in the bird study is slightly inaccurate. Therefore, adjustments of the bird data set are required before accurate comparisons can be made.

Two different preliminary butterfly data sets based on coarse distribution maps and confirmed records, respectively, were available for comparison. The coarse butterfly data set includes distributions for 939 taxa (Effendi, pers. comm.). Figure 10.12 is an overlay of the richness maps of butterflies and rattans and it illustrates high congruence in richness of rattans and butterflies in most of the northern and north-western parts of the island. The 26 grid cells selected as a priority set of areas for rattan conservation capture all butterfly taxa.

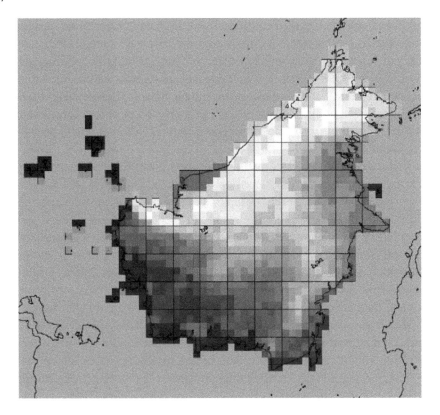

FIGURE 10.12 **(Colour Figure 10.12 follows p. 180.)** The congruence in richness patterns of rattans and birds and overlay of the richness maps of butterflies and rattans. White colours indicate grid cells of high rattan and bird richness. Green colours indicate that rattan diversity is relatively higher than butterfly diversity. Blue colours illustrate that bird diversity is relatively higher than rattan diversity.

The butterfly data set based on confirmed records includes 890 taxa (Effendi, pers. comm.). This data set is too preliminary for meaningful overlay with the rattan taxon richness map. The 26 priority rattan areas capture 787 (88.5%) butterfly taxa in this data set. The two preliminary comparisons between the rattan and butterfly data sets may indicate that the priority rattan areas to a great extent are also priority butterfly areas. However, both butterfly data sets must be improved before they can be used for accurate comparisons. It may be advisable to combine the two data sets.

The preliminary comparisons among rattans, birds and butterflies indicate that priority areas for rattan conservation may conserve much of the bird and butterfly fauna. However, due to the preliminary nature of this comparison, it is premature to draw any firm conclusions. Following this, there is currently no reason to believe that the patterns in rattan diversity and distribution are representative for wholesale biodiversity. Also, there is no reason to believe that the priority rattan areas are relevant as priority areas for wholesale biodiversity.

If the bird data set were adjusted and corrected, it would be interesting to conduct a thorough comparison of birds and rattans to identify a preliminary set of priority areas for biodiversity conservation on Borneo. As high-quality data suitable for WORLDMAP analysis become available also for butterflies and frogs, it would be very interesting to compile all data in an attempt to identify a set of priority areas for conservation on Borneo.

Other groups could be included in the analyses to make a priority set of areas for conservation on Borneo even better. Mammals are relatively well studied (Paine et al. 1985). The tree flora of Borneo is also subject of much scrutiny, especially in Sabah and Sarawak (Soepadmo and Wong 1995). Orchids (Orchidaceae) are equally well documented (Beaman et al. 2001; pers. comm.), and studies on the ginger flora (Zingiberaceae) of Borneo are ongoing (Poulsen, pers. comm.). Since insects probably account for more than half of the biodiversity in tropical forests (Groombridge and Jenkins 2002), it would probably also be advisable to include a few groups of insects.

In this study, available data and methods for area selection have been used to identify a set of priority areas for rattan conservation on Borneo. The rationale has been that basing rattan conservation priorities on available taxonomic data is the best and most feasible option, and it is firmly believed that the results of this study provide sound guidance for rattan conservation. In moving towards a set of priority areas for wholesale biodiversity conservation on Borneo, a similar approach may be applied — that is, to gather as much of the information available as possible for as many taxa as possible and to analyse the data to set priorities using methods based on complementarity. These priorities may be revised as more data become available for analysis. Interactive computer-based tools such as WORLDMAP allow for revision of data and for recalculating priorities. Given the current and projected rates of losses in biodiversity and natural landscapes in most parts of the tropics, awaiting additional biodiversity data before taking conservation action would be hazardous. It is necessary to use existing data on various taxonomic groups to set priorities and take conservation action now.

10.9   CONCLUSIONS

The key conclusions arising from this study on the rattans of Borneo are:

- Of the 144 rattan taxa living in Borneo, more than two thirds are endemic to the island and one quarter are narrowly distributed. Approximately 24% of all rattan species are found on Borneo.
- At least 18.8% of the rattans on Borneo and at least 23.5% of those endemic to the island are threatened. Three rattans are critically endangered.
- Patterns in taxon richness indicate that the northern and north-western areas contain more species than those in the south and the east.
- The patterns in range-size rarity indicate that endemism hotspots are found in the south-western Sarawak, Brunei and the central parts of Sabah.
- Well-protected grid cells contain 71.5% of the rattans found. The protection of rattans provided by the current set of reserves is not better than a random selection of reserves, and implementation of proposed reserves on Borneo would only marginally increase protection of rattans.
- Complementarity-based methods are the most efficient methods available to identify priority areas for rattan conservation on Borneo.
- Using complementarity, 23 of the 1087 grid cells in WORLDMAP are required to represent all rattan taxa in at least one grid cell. To represent all taxa at least once and, where possible, in a grid cell with more than 10% natural vegetation, 26 grid cells are required.

- The documentation of the rattan flora of Borneo provided in this study contributes to the implementation of the Global Strategy for Plant Conservation.
- Using studies that identify areas important to biodiversity as a tool to set conservation priorities will contribute to the implementation of the forest work programme of the Convention on Biological Diversity and the Plan on Implementation of the World Summit on Sustainable Development.

10.10   RECOMMENDATIONS

The findings in this study have resulted in the following key recommendations:

- Governments and others involved in rattan conservation on Borneo should consider the priority set of areas identified in this study in their work to protect rattans of Borneo.
- The governments on Borneo should engage in collaborative initiatives to ensure effective and efficient rattan conservation on Borneo.
- Those taxa that are threatened, particularly those critically endangered, should be subject to field studies that document their status. Urgent steps should be taken to secure threatened rattans where needed and where possible.
- The rattans on the Malay Peninsula should be studied with the aim of identifying priority areas for rattan conservation.
- The international biodiversity policy community should be advised to recommend application of complementarity-based area selection methods in conservation.
- A preliminary set of priority areas for biodiversity conservation on Borneo should be identified based on available distribution data for birds and rattans.
- A more comprehensive set of priority areas for biodiversity conservation on Borneo should be identified as high-quality data for other groups of organisms become available that allow for application of complementarity-based methods.

ACKNOWLEDGMENTS

In conducting this study, we have depended highly on the assistance, advice, hospitality and expertise of many people, especially Jakob Larsen who did much of the work for his MSc. It would need an essay to write all the affiliations and individual projects of each person involved so we have simply listed them here: Axel Dalberg Poulsen, Paul Williams, Jon Fjeldså, Bakhtiar Effendi, Michael Stabell, Januuarius Gobelik, Anthony Lamb, Jain Linton, Aninguh Surat, Hans Ulrik Skotte Møller, Waidi Sinun, Jens Kanstrup, Kho Ju Ming, Edward Alimin, Rasit bin Abdilah, Sulaiman bin Payau, Diwol Sundaling, Jim Sheh Ping, John Sugau, Markus Gumbili, Saw Leng Guan, John Beaman, Meesha Patel, Justin Moat and Don Kirkup. Finally, we gratefully thank WWF-Denmark, DANCED and SLUSE for the financial support that made this project possible.

REFERENCES

Andersen, J., Nilsson, C., De Richelieu, T., Fridriksdottir, H., Gobilik, J., Mertz, O., and Gausset, Q. (2001) Local use of forest products in Kuyongan, Sabah, Malaysia. In *A scientific journey through Borneo: The Crocker Range National Park Sabah*, vol. 2. *Socio-cultural and human dimension*, ed. I. Ghazally and A. Lamri. ASEAN Academic Press, London, 15–38.

Andersen, J., Boje, K.K. and Borchsenius, F. (in press) The palm flora of Tabin Wildlife Reserve. In *The Tabin Expedition II*, ed. M. Maryati. Universiti Malaysia Sabah, Kota Kinabalu.

Anonymous (1985) Country report Malaysia. *Proceedings of the rattan seminar*, 2–4 October 1984, Kuala Lumpur, Malaysia. The Rattan Information Centre.

Asmussen, C.B., Baker, W.J., and Dransfield, J. (2000) Phylogeny of the palm family (Arecaceae) based on RPS16 intron and TRNL-TRNF plastid DNA sequences. In *Monocots: Systematics and evolution*, ed. K.L. Wilson and D.A. Morrison. CSIRO, Melbourne, 525–537.

Baker, W.J. and Dransfield, J. (2000) Phylogeny, character evolution, and a new classification of the Calamoid palms. *Systematic Botany* 25: 297–322.

Balmford, A. and Gaston, K. (1999) Why biodiversity surveys are good value, *Nature*, 398: 203–204.

Beaman, T.E., Wood, J.J., Beaman, R.S., and Beaman, J.H. (2001) *Orchids of Sarawak*. Natural History Publications and the Royal Botanic Gardens, Kew, UK.

Blakeney, J. (2001) *Overview of forest law enforcement in East Malaysia*. WWF Malaysia.

Boje, K.K. (2000) Palms in Tabin Wildlife Reserve, Malaysia. Master thesis submitted to the Department of Systematic Botany, Institute of Biology, University of Aarhus, Denmark.

Borchsenius, F. and Skov, F. (1999) Conservation status of palms (Arecaceae) in Ecuador. *Acta Botanica Venezuela* 22: 221–236.

Christensen, H. (2002) *Ethnobotany of the Iban and the Kelabit*. A joint publication of Forest Department Sarawak, Malaysia; NEPCon, Denmark, and University of Aarhus, Denmark.

Convention on Biological Diversity (2002a) Global Strategy for Plant Conservation. Secretariat to the Convention on Biological Diversity (URL: http://www.biodiv.org/decisions/).

Convention on Biological Diversity (2002b) Forest Biological Diversity. Secretariat to the Convention on Biological Diversity (URL: http://www.biodiv.org/decisions/).

Csuti, B., Polaski, S., Williams, P., Pressey, R., Camm, J.D., Kershaw, M., Kiester, R., Downs, B., Hamilton, R., Huso, M., and Sahr, K. (1997) A comparison of reserve selection algorithms using data on terrestrial vertebrates in Oregon. *Biological Conservation* 80: 83–97.

Dawood, M.M. (1999) Potential use of Worldmap for exploring aspects of spatial patterns in biological data: An introduction to use of butterfly biodiversity assessment in Borneo. *Proceedings from the third entoma seminar*. Entomological Society of Malaysia, Malaysia.

Dransfield, J. (1979) *A manual to the rattans of the Malay Peninsula*. Malaysian Forest Records, no. 29, Kuala Lumpur.

Dransfield, J. (1981) Palms and Wallace's line. In *Wallace's line and plate tectonics*, ed. T.C. Whitmore. Clarendon Press, Oxford, 43–56.

Dransfield, J. (1984a) *The rattans of Sabah*. Forest Department, Sabah.

Dransfield, J. (1984b) The palm flora of Gunung Mulu National Park. In *Studies on the flora of Gunung Mulu National Park, Sarawak*, ed. A.C. Jermy. Forest Department, Sarawak, 41–75.

Dransfield, J. (1986) A guide to collecting palms. *Annals of Missouri Botanical Garden* 73: 166–176.

Dransfield, J. (1992) *The rattans of Sarawak*. Royal Botanic Gardens, Kew, and Sarawak Forest Department, Kuching.

Dransfield, J. (1997) *The rattans of Brunei Darussalam*. Ministry of Industry and Primary Resources, Brunei Darussalam.

Dransfield, J. and Johnson, D. (1990) The conservation status of palms in Sabah. *Malayan Naturalist* 43: 16–19.

Dransfield J. and Manokaran N. (1993) *Plant resources of South-East Asia No. 6: Rattans*. Pudoc Scientific Publishers, Wageningen.

Dransfield, J. and Patel, M. (2005) *Rattans of Borneo: An interactive key*. CD-ROM. Royal Botanic Gardens, Kew, UK.

FAO (2001) Global Forest Resources Assessment 2000, main report. Food and Agriculture Organization of the United Nations.

Fjeldså, J. (2000) The relevance of systematics in choosing priority areas for global conservation. *Environmental Conservation* 27: 67–75.

Fjeldså, J. and Rahbek, C. (1997) Species richness and endemism in South American birds: Implications for the design of networks of nature reserves. In *Tropical forest remnants*, ed. W.F. Laurance and R.O. Bierregaard, Jr. The University of Chicago Press, Chicago, 466–482.

Forest Watch Indonesia and Global Forest Watch. (2002) *The state of the forest: Indonesia*. Forest Watch Indonesia, Bogor, and Global Forest Watch, Washington.

Gait, B., Awang, R., and Urit, L.M. (1998) Checklist of commercial timbers and rattans in the Maliau Basin. In ed. M. Mohamed, W. Sinun, A. Anton, and Dalimin, 63–72.

Glowka, L., Burhenne-Guilmin, F., and Synge, H. (1994) *A guide to the convention on biological diversity*. IUCN, Gland.

Govaerts, R. and Dransfield, J. (2005) *World checklist of palms*. Royal Botanic Gardens, Kew, UK.

Greenpeace (2002) *The last of the world's ancient forests*. Greenpeace, Amsterdam.

Groombridge, B. and Jenkins, M.D. (2002) *World atlas on biodiversity. Earth's living resources in the 21st century*. University of California Press, Berkeley.

Holmgren, P.K., Holmgren, N.H., and Barnett, L.C. (1990) Index herbariorum. Part 1: The herbaria of the world. *Regnum Vegetabile* 120: 1–693.

Howard, P.C., Viscanic, P., Davenport, T.R.B., Kigenyi, F.W., Baltzer, M., Dickinson, C.J., Lwanga, J.S., Matthews, R.A., and Balmford, A. (1998) Complementarity and the use of indicator groups for reserve selection in Uganda. *Nature* 394: 472–475.

Humphries, C.J., Williams, P.H., and Vane-Wright, R.I. (1995) Measuring biodiversity value for conservation. *Annual Review of Ecology and Systematics* 26: 93–111.

ICPB (1992) *Putting biodiversity on the map: Priority areas for global conservation*. International Council for Bird Preservation, Cambridge.

Jandel Corporation (1993) *SigmaStat for Windows version 1.0*. Jandel Corporation.

Kiew R. and Pearce, K.G. (1990) Palms. In *The state of nature conservation in Malaysia*, ed. R. Kiew. Malayan Nature Society, Kuala Lumpur, 95–100.

Kirkup, D., Dransfield, J., and Sanderson, H. (1999) *The rattans of Brunei Darussalam — Interactive key on CD ROM*. Ministry of Industry and Primary Resources, Brunei Darussalam.

Lovett, J.C., Rudd, S., Taplin, J., and Frimodt-Møller, C. (2000) Patterns of plant diversity in Africa south of the Sahara and their implications for conservation management. *Biodiversity and Conservation* 9: 37–46.

Lund M.P. and Rahbek C. (2000) A quantitative biological analysis of the efficiency of Danish nature management — With emphasis on biological diversity (in Danish). *Arbejdspapir 2000*, 1. Det Økonomiske Råd, København.

Mace, G.M., Balmford, A., Boitani, L., Cowlishaw, G., Dobson, A.P., Faith, D.O., Gaston, K.J., Humphries, C.J., Vane-Wright, R.I., Williams, P.H., Lawton, J.H., Margules, C.R., May, R.M., Nicholls, A.O., Possingham, H.P., Rahbek, C., and Van Jaarsveld, A.S. 2000. It's time to work together and stop duplicating conservation efforts. *Nature* 405: 393.

Mackinnon, J., ed. (1997) *Protected areas systems review of the Indo-Malayan realm*. The Asian Bureau for Conservation and World Conservation Monitoring Centre, Canterbury, UK.

Margules, C.R., Nicholls, A.O., and Pressey, P.L. (1988) Selecting networks of reserves to maximize biological diversity. *Biological Conservation* 43: 63–76.

Microsoft (2000) Encarta Interactive World Atlas CD-ROM. Microsoft.

Minnemeyer, S. (2002) *An analysis of access into Central Africa's rainforests.* World Resources Institute, Washington, D.C.

Mittermeier, R.A., Myers, N., and Thomsen, J.B. (1998) Biodiversity hotspots and major wilderness areas: Approaches to setting conservation priorities. *Conservation Biology* 12: 516–520.

Myers, N. (1988) Threatened biotas: 'Hotspots' in tropical forests. *The Environmentalist* 8: 187–208.

Myers, N., Mittermeier, R.A., Mittermeier, C.G., Da Fonseca, G.A.B., and Kent, J. (2000) Biodiversity hotspots for conservation priorities. *Nature* 403: 853–858.

Noguerón, R. (2002) *Low-access forests and their level of protection in North America.* World Resources Institute, Washington, D.C.

Paine, J., Francis, C.M., and Phillipps, K. (1985) *A field guide to the mammals of Borneo.* The Sabah Society, Kota Kinabalu.

Pearce, K.G. (1989) Conservation status of palms in Sarawak. *Malayan Naturalist* 43: 20–36.

Pearce, K.G. (1994) The palms of Kubah National Park, Kuching division, Sarawak. *Malayan Nature Journal* 48: 1–36.

Prendergast, J.R., Quinn, R.M., Lawton, J.H., Eversham, B.C., and Gibbons, D.W. (1993) Rare species, the coincidence of diversity hotspots and conservation strategies. *Nature* 365: 335–337.

Pressey, R.L. (1994) Ad hoc reservations: Forward or backward steps in developing representative reserve systems? *Conservation Biology* 8: 662–668.

Pressey, R.L., Humphries, C.J., Margules, C.R., Vane-Wright, R.I., and Williams, P.H. (1993) Beyond opportunism: Key principles for systematic reserve selection. *Trends in Ecology and Evolution* 8: 124–128.

Pressey, R.L., Possingham, H.P., and Margules, C.R. (1996) Optimality in reserve selection algorithm: When does it matter and how much? *Biological Conservation* 76: 259–267.

Reid, W. (1998) Biodiversity hotspots. *Trends in Ecology and Evolution* 7: 275–280.

Rosen C. and Roberts L., eds. (2000) *World resources 2000–2001.* World Resources Institute, Washington, D.C.

Secretariat to the Convention on Biological Diversity (2001a) *Global biodiversity outlook.* Secretariat to the Convention on Biological Diversity, Montreal.

Secretariat to the Convention on Biological Diversity (2001b) *Handbook to the Convention on Biological Diversity.* Secretariat to the Convention on Biological Diversity, Montreal.

Soepadmo, E. and Wong, K.M., eds. (1995) *Tree flora of Sabah and Sarawak Volume One.* Sabah Forestry Department, Forest Research Institute and Sarawak Forestry Department, Malaysia.

SPSS 2002. *SigmaPlot 2002 for Windows version 8.01.* SPSS Inc.

Stabell, M. (2002) *En analyse af naturbeskyttelsesprioriteter for Borneos ynglefugle* (in Danish). Specialerapport, Zoologisk Museum, Københavns Universitet.

Uhl, N.W. and Dransfield, J. (1987) *Genera Palmarum.* Allen Press, Lawrence, KS.

Underhill, L.G. (1994) Optimal and suboptimæ reserve selection algorithms. *Biological Conservation* 70: 85–87.

United Nations (1992a) Report of the United Nations Conference on Environment and Development, Annex I: Rio declaration on environment and development. United Nations (URL:http://www.un.org/documents/ga/conf151/aconf15126-1annex1.htm).

United Nations (1992b) Earth Summit: Agenda 21. United Nations (URL:http://www.un.org/esa/sustdev/agenda21text.htm).

United Nations Forum on Forests (2002) Report of the second session. Economic and Social Council Official Records, 2002. Supplement No. 22. United Nations (URL:http://www.un.org/esa/forests/documents-unff.html).

Vane-Wright, R.I., Humphries, C.J., and Williams, P.H. (1991) What to protect? Systematics and the agony of choice. *Biological Conservation* 55: 235–254.

Williams, P.H. (1998) Key sites for conservation: Area-selection methods for biodiversity. In *Conservation in a changing world,* ed. G.M. Mace, A. Balmford, and J.R. Ginsberg. Cambridge University Press, Cambridge, 211–249.

Williams, P.H. (2001) *WORLDMAP version 4.20.15 (26.VIII.2001).* Privately distributed software, London (URL: http://www.nhm.ac.uk/research-curation/projects/worldmap/).

Williams, P.H. and Gaston, K. (1994) Measuring more of biodiversity: Can higher taxon richness predict wholesale species richness? *Biological Conservation* 67: 211–217.

Williams, P.H., Gibbons, D., Margules, C., Rebelo, A., Humphries, C., and Pressey, R. (1996) A comparison of richness hotspots, rarity hotspots and complementary areas for conserving diversity using British birds. *Conservation Biology* 10: 155–174.

Williams, P.H., Burgess, N.D., and Rahbek, C. (2000) Flagship species, ecological complementarity and conserving the diversity of mammals and birds in sub-Saharan Africa. *Animal Conservation* 3: 249–260.

Wong, K.M. (1998) Patterns of plant endemism and rarity in Borneo and the Malay peninsula. *Academica Sinica Monograph Series* 16: 139–169.

WWF and IUCN 1994. *Centres in plant diversity. Volume 1, Europe, Africa, South West Asia and the Middle East.* The World Conservation Union, Cambridge.

ENDNOTES

Endnote 1: The near-minimum set algorithm applied for a conservation goal of representing each species at least once (from Williams 2001):

1. Select all areas with species that have single records.
2. The following rules are applied repeatedly until all species are represented:
   A. Select areas with the greatest complementary richness in just the rarest species (ignoring less rare species); if there are ties, then:
   B. Select areas among ties with the greatest complementary richness in the next-rarest species and so on; if there are persistent ties, then:
   C. Select areas among ties with the greatest complementary richness in the next-next-rarest species and so on; if there are persistent ties, then:
   D. Select areas among ties with the greatest complementary richness in the next-next-next-rarest species and so on; if there are persistent ties, or no next- or next-next- or next-next-next-rarest species, then:
   E. Select areas among persistent ties with the lowest grid-cell number or at random (lowest grid-cell number may be used rather than random choice among ties in order to ensure repeatability in tests; other criteria, such as proximity to previously selected cells or number of records in surrounding cells, can be added). Repeat steps A through E until all species are represented.
3. Identify and reject any areas that, in hindsight, are unnecessary to represent all species.
4. Re-order areas by complementary richness.

Tools: the tools used for geographical referencing were, in order of priority:

1. **1**. A Sabah gazetteer; **2**. the Sarawak gazetteer included in Appendix 1 of Beaman et al. (2001)
2. **3**. Encarta Interactive World Atlas CD-ROM (Microsoft 2000)
3. **4**. GEOnet Names Server (http://164.214.2.59/gns/html/index.html)
4. **5**. Gazetteers developed by John Beaman (Royal Botanic Gardens, Kew) for Mt. Kinabalu (Beaman, pers. comm.)

Methods: according to the kind and quality of the locality description, the methods used for geographical referencing were, in order of priority:

1. A discrete and specific place name was a geographical reference found for this point (village, mountain top river mouth).
2. A linear location was given (watercourse as the geographical coordinates for the starting and ending or road), point was found and a simple average taken.
3. When a road locality was described by its distance to a discrete locality, such as a town, an interactive measuring tool was used (Microsoft 2000). An area was given (forest reserve, island), and the geographical coordinates for the extreme north, national park, south, east, and west of the area was found and the midpoint calculated.
4. The locality was marked on a small map Encarta Interactive World Atlas CD-ROM (Microsoft 2000) on the specimen's label (applied only to be used to determine the geographical coordinates of the specimens manually investigated at SAN).* This method was used only when there were no more than two grid cell borders between extreme sets of coordinates for a given area.

Documentation: the method and tools used for finding each geographical reference were recorded in the specimen database. If no method successfully revealed a geographical reference for a given specimen, it was excluded from the data set.

Endnote 2. The following terminology was used to describe the distribution of the rattans' taxa:

A. Occurrence of the rattan taxon on Borneo:
   1. >321 Grid cells = very widespread
   2. 33–320 Grid cells = widespread
   3. 9–32 Grid cells = local
   4. 1–8 Grid cells = narrow
B. Occurrence of the rattan taxa outside Borneo:
   1. If a rattan taxon is endemic to Borneo, this was included in the description.
   2. If a rattan taxon is not endemic to Borneo, the occurrence outside Borneo is described (based on Dransfield, pers. comm.).

Index

Systematics Association
Publications

1. Bibliography of Key Works for the Identification of the British Fauna and Flora, 3rd edition (1967)[†]
   Edited by G.J. Kerrich, R.D. Meikie and N. Tebble
2. Function and Taxonomic Importance (1959)[†]
   Edited by A.J. Cain
3. The Species Concept in Palaeontology (1956)[†]
   Edited by P.C. Sylvester-Bradley
4. Taxonomy and Geography (1962)[†]
   Edited by D. Nichols
5. Speciation in the Sea (1963)[†]
   Edited by J.P. Harding and N. Tebble
6. Phenetic and Phylogenetic Classification (1964)[†]
   Edited by V.H. Heywood and J. McNeill
7. Aspects of Tethyan biogeography (1967)[†]
   Edited by C.G. Adams and D.V. Ager
8. The Soil Ecosystem (1969)[†]
   Edited by H. Sheals
9. Organisms and Continents through Time (1973)[†]
   Edited by N.F. Hughes
10. Cladistics: A Practical Course in Systematics (1992)[*]
    P.L. Forey, C.J. Humphries, I.J. Kitching, R.W. Scotland, D.J. Siebert and D.M. Williams
11. Cladistics: The Theory and Practice of Parsimony Analysis (2nd edition)(1998)[*]
    I.J. Kitching, P.L. Forey, C.J. Humphries and D.M. Williams

[*] Published by Oxford University Press for the Systematics Association

[†] Published by the Association (out of print)

SYSTEMATICS ASSOCIATION SPECIAL VOLUMES

1. The New Systematics (1940)
   Edited by J.S. Huxley (reprinted 1971)
2. Chemotaxonomy and Serotaxonomy (1968)[*]
   Edited by J.C. Hawkes
3. Data Processing in Biology and Geology (1971)[*]
   Edited by J.L. Cutbill

4. Scanning Electron Microscopy (1971)*
   Edited by V.H. Heywood

5. Taxonomy and Ecology (1973)*
   Edited by V.H. Heywood

6. The Changing Flora and Fauna of Britain (1974)*
   Edited by D.L. Hawksworth

7. Biological Identification with Computers (1975)*
   Edited by R.J. Pankhurst

8. Lichenology: Progress and Problems (1976)*
   Edited by D.H. Brown, D.L. Hawksworth and R.H. Bailey

9. Key Works to the Fauna and Flora of the British Isles and Northwestern Europe, 4th edition (1978)*
   Edited by G.J. Kerrich, D.L. Hawksworth and R.W. Sims

10. Modern Approaches to the Taxonomy of Red and Brown Algae (1978)
    Edited by D.E.G. Irvine and J.H. Price

11. Biology and Systematics of Colonial Organisms (1979)*
    Edited by C. Larwood and B.R. Rosen

12. The Origin of Major Invertebrate Groups (1979)*
    Edited by M.R. House

13. Advances in Bryozoology (1979)*
    Edited by G.P. Larwood and M.B. Abbott

14. Bryophyte Systematics (1979)*
    Edited by G.C.S. Clarke and J.G. Duckett

15. The Terrestrial Environment and the Origin of Land Vertebrates (1980)
    Edited by A.L. Pachen

16 Chemosystematics: Principles and Practice (1980)*
    Edited by F.A. Bisby, J.G. Vaughan and C.A. Wright

17. The Shore Environment: Methods and Ecosystems (2 volumes) (1980)*
    Edited by J.H. Price, D.E.C. Irvine and W.F. Farnham

18. The Ammonoidea (1981)*
    Edited by M.R. House and J.R. Senior

19. Biosystematics of Social Insects (1981)*
    Edited by P.E. House and J.-L. Clement

20. Genome Evolution (1982)*
    Edited by G.A. Dover and R.B. Flavell

21. Problems of Phylogenetic Reconstruction (1982)
    Edited by K.A. Joysey and A.E. Friday

22. Concepts in Nematode Systematics (1983)*
    Edited by A.R. Stone, H.M. Platt and L.F. Khalil

23. Evolution, Time and Space: The Emergence of the Biosphere (1983)*
    Edited by R.W. Sims, J.H. Price and P.E.S. Whalley

24. Protein Polymorphism: Adaptive and Taxonomic Significance (1983)*
    Edited by G.S. Oxford and D. Rollinson

25. Current Concepts in Plant Taxonomy (1983)*
    Edited by V.H. Heywood and D.M. Moore

26. Databases in Systematics (1984)*
    Edited by R. Allkin and F.A. Bisby

27. Systematics of the Green Algae (1984)*
    Edited by D.E.G. Irvine and D.M. John

28. The Origins and Relationships of Lower Invertebrates (1985)[‡]
    Edited by S. Conway Morris, J.D. George, R. Gibson and H.M. Platt

29. Infraspecific Classification of Wild and Cultivated Plants (1986)[‡]
    Edited by B.T. Styles

30. Biomineralization in Lower Plants and Animals (1986)[‡]
    Edited by B.S.C. Leadbeater and R. Riding

31. Systematic and Taxonomic Approaches in Palaeobotany (1986)[‡]
    Edited by R.A. Spicer and B.A. Thomas

32. Coevolution and Systematics (1986)[‡]
    Edited by A.R. Stone and D.L. Hawksworth

33. Key Works to the Fauna and Flora of the British Isles and Northwestern Europe, 5th edition (1988)[‡]
    Edited by R.W. Sims, P. Freeman and D.L. Hawksworth

34. Extinction and Survival in the Fossil Record (1988)[‡]
    Edited by G.P. Larwood

35. The Phylogeny and Classification of the Tetrapods (2 volumes)(1988)[‡]
    Edited by M.J. Benton

36. Prospects in Systematics (1988)[‡]
    Edited by J.L. Hawksworth

37. Biosystematics of Haematophagous Insects (1988)[‡]
    Edited by M.W. Service

38. The Chromophyte Algae: Problems and Perspective (1989)[‡]
    Edited by J.C. Green, B.S.C. Leadbeater and W.L. Diver

39. Electrophoretic Studies on Agricultural Pests (1989)[‡]
    Edited by H.D. Loxdale and J. den Hollander

40. Evolution, Systematics, and Fossil History of the Hamamelidae (2 volumes)(1989)[‡]
    Edited by P.R. Crane and S. Blackmore

41. Scanning Electron Microscopy in Taxonomy and Functional Morphology (1990)[‡]
    Edited by D. Claugher

42. Major Evolutionary Radiations (1990)[‡]
    Edited by P.D. Taylor and G.P. Larwood

43. Tropical Lichens: Their Systematics, Conservation and Ecology (1991)[‡]
    Edited by G.J. Galloway

44. Pollen and Spores: Patterns and Diversification (1991)[‡]
    Edited by S. Blackmore and S.H. Barnes

45. The Biology of Free-Living Heterotrophic Flagellates (1991)[‡]
    Edited by D.J. Patterson and J. Larsen

46. Plant–Animal Interactions in the Marine Benthos (1992)[‡]
    Edited by D.M. John, S.J. Hawkins and J.H. Price

47. The Ammonoidea: Environment, Ecology and Evolutionary Change (1993)[‡]
    Edited by M.R. House

48. Designs for a Global Plant Species Information System (1993)‡
    Edited by F.A. Bisby, G.F. Russell and R.J. Pankhurst

49. Plant Galls: Organisms, Interactions, Populations (1994)‡
    Edited by M.A.J. Williams

50. Systematics and Conservation Evaluation (1994)‡
    Edited by P.L. Forey, C.J. Humphries and R.I. Vane-Wright

51. The Haptophyte Algae (1994)‡
    Edited by J.C. Green and B.S.C. Leadbeater

52. Models in Phylogeny Reconstruction (1994)‡
    Edited by R. Scotland, D.I. Siebert and D.M. Williams

53. The Ecology of Agricultural Pests: Biochemical Approaches (1996)**
    Edited by W.O.C. Symondson and J.E. Liddell

54. Species: the Units of Diversity (1997)**
    Edited by M.F. Claridge, H.A. Dawah and M.R. Wilson

55. Arthropod Relationships (1998)**
    Edited by R.A. Fortey and R.H. Thomas

56. Evolutionary Relationships among Protozoa (1998)**
    Edited by G.H. Coombs, K. Vickerman, M.A. Sleigh and A. Warren

57. Molecular Systematics and Plant Evolution (1999)
    Edited by P.M. Hollingsworth, R.M. Bateman and R.J. Gornall

58. Homology and Systematics (2000)
    Edited by R. Scotland and R.T. Pennington

59. The Flagellates: Unity, Diversity and Evolution (2000)
    Edited by B.S.C. Leadbeater and J.C. Green

60. Interrelationships of the Platyhelminthes (2001)
    Edited by D.T.J. Littlewood and R.A. Bray

61. Major Events in Early Vertebrate Evolution (2001)
    Edited by P.E. Ahlberg

62. The Changing Wildlife of Great Britain and Ireland (2001)
    Edited by D.L. Hawksworth

63. Brachiopods Past and Present (2001)
    Edited by H. Brunton, L.R.M. Cocks and S.L. Long

64. Morphology, Shape and Phylogeny (2002)
    Edited by N. MacLeod and P.L. Forey

65. Developmental Genetics and Plant Evolution (2002)
    Edited by Q.C.B. Cronk, R.M. Bateman and J.A. Hawkins

66. Telling the Evolutionary Time: Molecular Clocks and the Fossil Record (2003)
    Edited by P.C.J. Donoghue and M.P. Smith

67. Milestones in Systematics (2004)
    Edited by D.M. Williams and P.L. Forey

68. Organelles, Genomes and Eukaryote Phylogeny (2004)
    Edited by R.P. Hirt and D.S. Horner

69. Neotropical Savannas and Seasonally Dry Forests: Plant Diversity, Biogeography and Conservation (2006)
    Edited by R.T. Pennington, G.P. Lewis and J.A. Rattan

70. Biogeography in a Changing World (2006)
    Edited by M.C. Ebach and R.S. Tangney

71. Pleurocarpous Mosses: Systematics & Evolution (2006)
    Edited by A.E. Newton and R.S. Tangney

72. Reconstructing the Tree of Life: Taxonomy and Systematics of Species Rich Taxa (2006)
    Edited by T.R. Hodkinson and J.A.N. Parnell

* Published by Academic Press for the Systematics Association
† Published by the Palaeontological Association in conjunction with Systematics Association
‡ Published by the Oxford University Press for the Systematics Association
** Published by Chapman & Hall for the Systematics Association

9 780415 332903